Student Study Guide

ANN RATCLIFFE
University of Northern Colorado

DAVID F. DEVER
Macon State College (retired)

STANDA

The Extraordinary Chemistry of Ordinary Things

Fourth Edition

CARL H. SNYDER
University of Miami

WILEY

Cover Photo: Steve Lupton/CORBIS.

To order books or for customer service call 1-800-CALL-WILEY (225-5945).

ISBN 0-471-42359-9

Printed in the United States of America

10 9 8 7 6 5 4 3 2 1

Printed and bound by Bradford & Bigelow, Inc.

About The Authors

Ann Ratcliffe received the A.B. degree in chemistry from Randolph-Macon Woman's College and the Master of Arts in Teaching from Duke University. She taught high school chemistry in both public and private schools, including four years at the American Community School in Beirut, Lebanon. At Oklahoma State University she coordinated the general chemistry teaching laboratories and taught in the general chemistry program. Currently she is Program Coordinator for the Center for Precollegiate Studies and Outreach at the University of Northern Colorado. She is the author of *Chemistry: The Experience*, a lab manual for non-science majors which employs the guided inquiry method of learning (John Wiley & Sons, Inc., 1993).

David F. Dever received his bachelor's degree at Spring Hill College, master's degree from The Florida State University, and was a graduate school colleague of Carl H. Snyder (the author of *The Extraordinary Chemistry of Ordinary Things*) while they were both earning their PhDs at The Ohio State University in the late fifties. He worked at the Los Alamos Scientific Laboratory as a research fellow, the U.S. Bureau of Mines as a project leader in air-pollution research, and helped start an engineering college on the Persian Gulf for the Saudi Arab Government. He taught 34 years at Macon State College and is now retired.

Preface to the Fourth Edition

In the fourth edition of this *Guide* we continue our emphasis on developing good study habits. Throughout the first edition Dave Dever provided study tips gleaned from his research on how to study. Based on Heiman and Slomianko s *Learning to Learn*, the first three rules will get you started.[*]

- **Sit in front of the class**. This sounds trite, but it turns out to be very important. If you think this is not so, sit up front, notice who sits in the back, and see which group thins out by midterm.

- **Ask lots of questions.** At first reading, this advice seems to mean Ask the professor. However, as you progress through this *Guide,* you will see that the advice goes even further. Ask yourself questions as you listen in class or read the chapter beforehand; more of this below.

- **Keep your *eyes* on the action.** Look at the instructors when they are talking or writing on the board; look at your fellow students when they are asking questions. *Discretion and good judgment are required here*: if you are supposed to look at a graph or table in a book, do so. In all cases, stay in touch with what is going on, *all the time*. (And it s easier to do this if you are sitting in the front of the class!)

Making It Work

An effective way to get the most from a class is to form a study group. Organize a small group of students in your class with whom you can meet regularly to study. This can be students you carpool with, students at your lab bench, or friends you know in the class. Meet after each class to compare notes, ask questions, write questions for each other to contemplate, work problems, explain problems to each other.

One of the biggest benefits to accrue from this practice is seeing the way other students deal with the course; you may see that there is another way to do a numerical problem, or that there is another angle to be developed on an open-ended question. The biggest benefit, however, is the fact that, among your group, 80% of the questions the professor would think of asking have already passed before your eyes. If your group isn t active, find another one. It s that important!

About this Guide

As you begin using this *Guide* you will notice that in each chapter an introductory scenario is followed by a Chapter Overview putting forth the basic principles addressed in the respective chapter in *The Extraordinary Chemistry of Ordinary Things* (ECOT). Then certain concepts are amplified in the section titled For Emphasis, where you will find sample problems, perhaps a different way to look at a concept in ECOT, and basic helpful hints. Following that, Questions Answered provides solutions to the Questions at the end of each section in the chapter in ECOT. This is followed by a few Supplementary Exercises to give you yet more practice. Solutions to the Supplementary Exercises are in the Appendix of this *Guide* . At the end of each chapter is the Where You Might Goof Up . . . section. Take this section seriously, because it highlights those things in the ECOT chapter that just might trip you up.

Acknowledgments

I am indebted to Nathan Barrows, doctoral student in chemical education at the University of Northern Colorado, for invaluable help with this edition. And my thanks go also to my editor, Jennifer Yee, with whom it has once again been a pleasure to work.

Ann Ratcliffe

---

[*] Marsha Heiman and Joshua Slomianko, *Learning to Learn*, Learning to Learn, Inc., 28 Penniman Road, Allston, M.A 02134.

Contents

CHAPTER 1

An Introduction to Chemistry

TURNING ON THE LIGHT

Years ago in the well-baby nursery of the hospital of a town in upstate New York, healthy newborns suddenly began to show distress: they took less milk, cried extensively, showed puffy skin and in a few cases produced bloody urine. Since the subjects were confined to one room and were attended by only a few nurses who passed among all of them, an early suspect was sepsis: bacteria in the blood stream. This contagious condition would explain the constellation of symptoms. The time spent excluding this hypothesis was very expensive: a few of the babies died. When the search was widened (both clinically and physically) to include the kitchen where the formulas were prepared, the causes were found. The sugar and salt bins were labeled on the container lids rather than the sides of the containers themselves, and one of the workers had accidentally exchanged the lids. Thus, when a formula called for so much sugar, the infant got salt instead; this gravely disrupted the electrolyte balance of the child and produced the symptoms described above. Salt and sugar look alike, have approximately the same density, pour the same, so who would know? How could two common, easily confused compounds be so diverse? Act so differently?

One way to learn effectively is to ask a lot of questions. Notice that the paragraph telling the true story above ended with some questions. As you read a sentence or at most a paragraph, you should stop and "quiz the book," asking questions based on what you have just read. The questions stated above are reasonable ones, but not the only plausible ones. There are no "correct" questions in these exercises, just good ones.

Even though you may have already read the first chapter of ECOT, go back and read it again. While you may be new at this "quizzing the book" technique, force yourself to stop periodically to ask a paragraph a few questions. After you have been doing this for awhile it will become automatic, but for the first few times, make yourself stop and sieve your mind for questions that want to bob to the surface. Two questions that come to the minds of some students who have read the first paragraphs of ECOT are:

- What is another way to tell salt and sugar apart than by taste?

- What is the difference in electrical behavior of the two solutions, and how do you measure this behavior?

Naturally, there are other equally valid questions to be asked that would come to mind. Whatever they are, *write those questions in the margin of the page*. As it turns out, the questions above are answered in the next few paragraphs which describe the conductivity experiment.

Continue this process through the first chapter of ECOT and see how much different your understanding of the material is than you expected it to be. Some students taking their first chemistry course have already done what is recommended in the paragraph above, and their most-often listed questions are at the end of the "For Emphasis" section on p. 5. After you have listed your questions, why not give theirs a look?

Chapter Overview

- This chapter is an introduction to the subject of science in general and chemistry in particular.

- An experiment (demonstration, student-performed or -thought) is described concerning solutions made from substances that look alike.

- The solutions' electrical behaviors differ.

- Atoms, molecules and ions are introduced.

- The distinction between electrolytes and nonelectrolytes is made.

- Chemistry is defined and three practical points are made:

    1. chemical reactions are the main source of energy both for our bodies and our planet at large;

    2. anything material that exists is made up of chemicals, including us;

    3. chemicals themselves are neither good nor bad—it's how we use them that matters. (Nitroglycerine is a most violent explosive, but it is also a heart medicine.)

- There is an important property of all compounds: CONSTANT COMPOSITION.

- An explanation is offered for the difference in behavior of the two solutions.

- The nature and process of science is described and points are emphasized, some of which might surprise some nonscientists. For example, "No theory is true."

- This process of science, "The Scientific Method," consists of observation, careful collection of data, hypothesis, and plausible explanation of the data (facts), this last of which we call theory.

- Science is knowledge that we obtain by the systematic study of the universe; technology is the application of science to our lives. For example, chemical and electrical properties of the element silicon, Si, have been *studied* over many decades by careful experimentation and then *applied* to produce the computer chip, giving us what we refer to as "computer technology."

For Emphasis
Elements, Compounds and Mixtures

In this first chapter of ECOT, science is seen in more detail than in the science-fiction movie or newspaper report. Scientists deal with both facts *and* ideas. In this chapter, the facts are whether the light goes on or not. The ideas are the explanation of *why* the light goes on or not. The ideas in this case are called *theory*. Scientists frequently use the words *theory, idea, model* or even *explanation* interchangeably. When you hear someone say, "I don't like science; it's just a bunch of facts to be memorized and that's a pain," realize that someone has been poorly schooled. Otherwise, what would scientists have to argue with each other about? If all they could do was criticize another's data rather than the ideas, what fun is that?

One of the ways chemists categorize material is by calling them **mixtures**, **compounds** or **elements**. A **mixture** can have varying composition, such as:

- a 5% sugar solution or a 7% sugar solution is still a mixture of sugar and water;

- a wheelbarrow of concrete mix can have one, two or three shovels of sand in it along with the cement, gravel and water, but it is still concrete;

- some other examples of mixtures: gasoline, rubbing alcohol, stainless steel and the paper this is written on.

A **compound**, as discussed in the first chapter of ECOT, has constant composition; that is to say, no matter how many different kinds of materials are in the compound, the *relative* amounts of them are fixed. Some examples:

- as discussed in ECOT, sodium chloride is formed from the chemical reaction between sodium and chlorine and always is formed in the ratio of 23 g of Na to 35.5 g of Cl, or 23/35.5 = 0.648. It does not matter whether the salt comes from a salt-flat in Montana or is isolated from water in Vladivostok harbor;

- water is a compound and is always found in the ratio of 2 g of hydrogen to 16 g of oxygen whether it is sampled from the polar ice cap or from the vapor exhaled from our lungs;

- carbon dioxide, whether exhaled from our lungs or expelled from an auto's tailpipe, is always found in the ratio of 12 g of carbon to 32 g of oxygen;

- some other examples: ethyl alcohol (alcohol), sugar (of course), vitamin C (ascorbic acid), aspirin (acetylsalicylic acid), nicotine, and sodium bicarbonate (baking soda).

An **element** is the simplest item a chemist deals with. It is always 100% of a single substance: carbon is 100% carbon atoms, sodium is 100% sodium atoms and iron is 100% iron atoms. In no case is a pure element mixed with anything else. A list of examples is found in the Periodic Table inside the front cover of this and most other physical science books. This makes life easy when you are asked whether or not a substance is an element: if it can be found on the Table, it is an element.

The Scientific Method

The scientific method discussed in ECOT is easy to understand, appreciate and remember if it is kept in mind that the steps in this method going from start to finish become larger and larger things to think about;

- **observation**: Just looking at what is happening? Not quite. There is a big difference between looking at something that happened and watching an event. If possible, careful records should be made, photos taken, etc. The student sees that a piece of 2x4 floats in water.

- **experiment**: This can be thought of as an "arranged" observation. The student tries to duplicate (or improve on, but at least control) the conditions of the initial observation. More data are taken. The student saws the 2x4 into two halves; they float as well. As experiments are repeated, are there any regularities noticed? Any generalities to be made? If so, the next step is in order.

- **hypothesis**: The generality obvious to the student is: objects with sharp corners and right-angle corners float. The student goes to garage sales and picks up many other items that the hypothesis predicts will float; it is a disappointment to find out that sharp-edged glass, iron and granite do not float. The student looks for other items not necessarily having sharp edges that might float. Round wooden balls, sticks, pegs, and other wooden items are found to float; bricks, stones, brass door keys and lead do not. Another hypothesis is cast: wooden items float. The student looks far and wide, but is unable to find any item made of wood, no matter what the shape, that does not float in water. After extensive data supporting the hypothesis is examined, the student is ready to make another kind of generalization.

- **law**: All items made of wood float. A law is a very powerful way of thinking. Should a list of all the items which float be carried around in the head? Folded neatly and placed in a pocket? Of course not. The regularity that the student noticed is a much more general method of packaging facts. Law: a general way to state a set of data. Intellectually, the student is ready to move on.

- **theory**: A plausible explanation of the facts. Why does wood float in water? The student notices that wood feels "light". An explanation is proposed: Wood floats in water because it is lighter than water. It sounds like a good explanation so the student prepares to write friends and acquaintances to spread the news. While this is being done, a world-traveling cousin of the student returns home with a throw-stick purchased from an Australian aborigine; it is used for hunting in the Outback and is made of ironwood. Being the first to hear of this, the cousin fills the sink with water and places the throw-stick on the water. It sinks. This theory did what every good theory does: it provokes challenges. In this particular case, the theory could not withstand the challenge, so it must be either modified or discarded (in which case, the student must start over with an entirely new explanation).

Thinking With Numbers

If you are comfortable with numbers and simple algebra, skip this section. If not, read on.

No matter what you are doing in your daily life (whether you are studying chemistry or not), at some time in every day you use numbers to solve a problem. If the price of a gallon of gasoline is $0.95, you know that for $5 you can get about 5 gallons of gas. If apples are 3 for $1, 6 apples will cost $2.

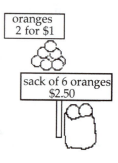

oranges
2 for $1

sack of 6 oranges
$2.50

Now let's say you go to the grocery store to buy 6 oranges, and you find that on one display individual oranges are 2 for $1 and on another there is a sack of 6 oranges for $2.50. Before you look at the solution in the box below, *in your head* decide which oranges to buy.

Solution: In seconds you know that the sack of 6 oranges is the bargain. It will cost you less per orange than buying them individually. There are several ways you might have approached the problem in your head. You might have said to yourself, "For 2 oranges I have to spend a dollar, so for 6 oranges (3 x 2) I will have to spend $3.00. Therefore it's cheaper to buy the sack of oranges." Or, if you like the challenge of playing with numbers, you might have run through it in your head like this:

If two oranges cost $1, then one orange costs $\left(\dfrac{\$1.00}{2}\right)$, or $0.50.

Six oranges, then, will cost $6\left(\dfrac{\$1.00}{2}\right)$, or 6($0.50), which is $3.00.

That you do this kind of mental calculation every day means, of course, that you are good at handling ratios and proportions in your head. Your math education in secondary school has paid off, and you can apply this technique to any problem that requires it, even chemistry.

In this particular example the ratio of dollars to individual oranges is 1 dollar to 2 oranges. Of course the reciprocal relationship makes a true statement, as well, i. e., 2 oranges to 1 dollar. You will see as we are solving chemistry problems together that simple ratios like these will describe chemical relationships, like that of hydrogen to oxygen in water or of sodium to chlorine in sodium chloride.

Quizzing the Chapter

Read this section *after* you have read and quizzed Chapter 1 yourself.

Twenty-five students, about half of them chemically naive, read Chapter 1 of ECOT and were asked, every few paragraphs, to write down questions provoked by the text. The list below gives the questions asked with the highest frequency by those students. At the beginning of this chapter, you were asked to read Chapter 1 of ECOT and write down questions prompted by the reading. Compare your list with this one.

- What is the chemical name for sugar?
- What is an electrical current?
- How do sodium chloride and sucrose act in water?
- What is a crystalline solid?
- Why is the electrical behavior of the two in water different?
- What is a filament?
- How can a flashlight tell the difference between sugar and salt?
- Why warm water *vs.* cold?
- Why does salt conduct and sugar not?
- Why is water a poor conductor?
- Why are the electrons pushed too weakly?
- What is ordinary matter?
- What are atoms, molecules and chemical bonds?
- What determines whether a chemical is good or bad?
- Why doesn't sugar conduct electricity?
- What is an electrolyte?
- What is a polymeric plastic?
- What is atmospheric ozone?
- What is a transmutation?
- Why can't elements be changed?
- What is a ratio?
- What is the scientific method?
- Is a theory true?

It would be a surprise not to find substantial overlap between your list of questions and the ones above. You will have noticed that you might have written down a question that was answered a few lines, a few paragraphs later or on the next page. Good. The fact or explanation meant more to you when you came to it, because you had asked about the item before running into it. You can see how this applies to disciplines other than chemistry even though chemistry is what we are treating here.

Questions Answered

1.1 The solution would conduct electricity, because ions are dispersed throughout the solution. The sugar molecules, also dispersed throughout the solution, don't conduct electricity.

1.2 a. Plants must be watered regularly or they will die. Humans are no different; we can survive without water for only a matter of days. b. If water enters a person's lungs, it can cause death by drowning.

1.3 No. In the experiment at the beginning of the chapter, water did not conduct electricity. (Remember: Chemistry is an *experimental* science!)

1.4 This problem is similar to the one in the Example on page 10 in ECOT. The small mass of sodium, 5 g, will require less than 15 g of chlorine.

$$5.0 \text{g sodium} \left(\frac{35.5 \text{g chlorine}}{23.0 \text{g sodium}} \right) = 7.7 \text{g chlorine combined with sodium}$$

Therefore, 5.0 g of sodium are reacting with 7.7 g of chlorine to produce 12.7 g of sodium chloride. The quantity of chlorine remaining would be: 15 g − 7.7 g = 7.3 g chlorine.

1.5 Dissolve the aspirin tablet in water and have someone test the conductivity of the resulting solution. If the light bulb glows, the solution contains ions that must have come from the aspirin tablet, because pure water contains too few ions to cause the bulb to glow.

Supplementary Exercises

S1. Elements lithium and fluorine form a compound, lithium fluoride. If 19.0 g of fluorine and 6.9 g of lithium react with each other and there is no excess of either one left over, how many g of lithium are needed to react with 12 g of fluorine?

S2. a. Elements potassium and bromine form a compound, potassium bromide. Experiment shows that 39.1 g of potassium always react with 79.9 g of bromine. Suppose you start with 10.0 g of potassium and in one experiment react it with 0.1 g of bromine; in the next experiment, you react it (10.0 g again) with 0.2 g bromine; in the next, you react it with 0.3 g bromine and so on. Each time you run another experiment, you increase the amount of bromine by 0.1 g. Describe qualitatively how the products of each individual reaction change as the experiment progresses from 0.1 g bromine to 5 g bromine.

b.   What amount of bromine reacts with 10.0 g of potassium?

S3.   Calcium and sulfur react with each other in the ratio of 40.1 g of calcium to 32.1 g of sulfur to form the compound calcium sulfide. If 15.0 g of calcium are reacted with 15 g of sulfur, how much sulfur will be left unreacted?

S4.   Calcium and chlorine form calcium chloride in the following ratio:

$$\left(\frac{40.1 \text{ g of calcium}}{70.0 \text{ g of chlorine}} \right)$$

If 50.0 g of calcium reacts with 100.0 g of chlorine, how much of which element is left over?

Where You Might Goof Up...

- Forgetting the definitions of *matter* and *chemistry*.

- Failing to distinguish among *atoms* and *ions*, *cations* and *anions*, *mixtures*, *compounds* and *elements*.

- Assuming that after a theory is proposed and many scientists think it's "pretty good," they all go on to something else and the theory is left poured in concrete. NOT SO. Theories are never safe; they are constantly under challenge; they are never the last word.

- Forgetting that ions can be at rest in a solid as well as in motion in solution carrying an electrical charge.

- Failing to realize that although most any student can perform a straightforward experiment, the scientific act is not complete until the results are interpreted. The more complete the exposition, the more useful to the science in particular and society in general.

- Failing to understand the difference between *science* and *technology*.

- Forgetting that chemistry is an *experimental* science.

CHAPTER 2

Atoms
and Elements

THE BUILDING BLOCKS
OF CHEMISTRY

The search for an understanding of the fundamental nature of existence is as old as humankind. From an early Indian philosophy that all existence is solidified water, to the Greek philosopher Empedocles who suggested four basic elements, earth, air, fire, and water, to the Greek atomist Democritus, whom we met in ECOT, people have looked for ways to understand the Universe. To the ancient peoples, however, this was a philosophical, not an experimental, activity.

The need for goods, such as glass, pottery, bronze, and dyes, gave rise to industries in the beginnings of human activity, but the discovery of methods of production was largely accidental, passed down from generation to generation. If one had been able to ask an ancient metalworker, say 5000 years ago, what was in the bronze he was producing, he might have answered, "Some black rock, some white rock, and fire." The woman producing dyes for her village knew which herbs to pick to produce the subtle earth colors her tribe used, but she had no idea of the chemical nature of the dye.

It was in the beginning of the Middle Ages, around the 5th or 6th century, that the earliest experimentalists began the search that came to be known as "alchemy," a search not only for a method of transmuting lead into gold (impossible by chemical means, we now know), but for the mystical nature of all existence. Alchemy evolved over time to become a much more systematic study of the chemistry of nature and much less mystical. In the 17th century Robert Boyle (we'll meet him again in the gas laws in Chapter 12) defined the concept of "element" that we understand today, and in the early 19th century John Dalton (who'll also appear again in Chapter 12) wrote an atomic theory that, for the most part, has survived intact and is the basis for the atomic theory in our late 20th century textbooks, including ECOT.

Chapter Overview

- A demonstration with paper clips is described. Give it a try.

- The idea of *atoms* is developed.

- The concept of the infinitesimal size of an atom is considered.

- A distinction is made between *mass* and *weight*.

- An extension of the ancient Greek idea of an atom (unbreakable, indivisible) is introduced.

- The Greek concept of the atom is expanded upon, and the parts of the atom are described: *protons*, *neutrons* and *electrons*.

- How to specify the exact identity of an atom using *mass number*, *atomic number*, and the *symbol* of an element is presented.

- *Isotopes*, atoms with different mass but the same atomic number and chemical behavior are introduced and described.

- Bigger and bigger atoms are constructed by packing more and more of the atomic parts together.

- The difference between adding a neutron to a nucleus and adding a proton to a nucleus is stressed.

- Electron structures are described.

For Emphasis

Mass and Weight

One of the ways to think about the difference between *mass* and *weight* is to ask the question, "If astronauts are weightless in a space station, how does the doctor on board keep such careful track of their weight loss and gain during the trip?" If there is no apparent "pull" by the Earth and no "push" on, say, a bathroom scale, how are the data collected?

The doctor uses what is called a Hooke's Law apparatus that measures a change in mass even though there is no apparent weight change to be assessed. The apparatus works like this: A way that 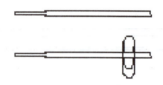 one of the authors (DFD) used to annoy his grade school teachers was to hang a nail file off the edge of his school desk, hold it tightly in place and pluck the end protruding from the edge of the desk. It made a healthy "twang" and resounded throughout the room. (It could also be slipped into a sock without attracting attention.) Try the same thing a few times and note the pitch of the twang. Then take one of the paper clips from the demonstration you did at the beginning of the ECOT chapter and slip it sideways over the end of the file hanging over the edge of the desk; it will be easiest if the clip makes a "T" with the file. Making sure that the file is extended the same length out from the desk, pluck it again a few times. If you have done everything right, you should notice that the increase in the *mass* of the overhanging section of the file has led to a lowering of the pitch of the sound. Add another paper clip if it will fit and try it again. Without moving the file, detach the paper clips, pluck the file and note the rise in the pitch.

This works just as well in outer space as it does on the earth. The astronaut is attached to a brace mounted on a flexible bar and the bar is plucked. The pitch (frequency) with which the bar oscillates tells the doctor the mass of the astronaut. Although the *weightless* astronaut probably floated into the brace to begin with, the doctor could tell if there had been any gain or loss in *mass* from, say, two days before. An increase in pitch means that there is a loss, a decrease in pitch means that there is a gain. A careful measure of the pitch will indicate how much gain or loss. The mass is measured by the attempt of the brace (nail file) to restore itself to the original position before it was distorted from its equilibrium (rest) position. For the idea of weight to mean anything, there must be a comparison between two bodies statically interacting with each other, an attraction between two bodies. One can talk about the mass of a body without comparing it to another; mass is an intrinsic property of a body itself. The brace (+ astronaut) in the space-station and the nail file (+ paper clip) changed pitch because they changed mass; outside forces had nothing to do with it.

Mass and Weight in ECOT

In Section 2.3, the author of ECOT notes that we tend to use *mass* and *weight* interchangeably, and he explains why this is acceptable on the earth's surface. But, he also points out that this doesn't mean that it is not important to understand the difference between them. You will see in the problems in Chapter 2 that "weight" is used more often than is "mass." One reason for this is that we have no verb "to mass" in English, as we have "to weigh." So we go to a balance to determine the mass of a product, but the lab manual might say "weigh the product." So you can see why there is confusion! Avoid your own confusion by remembering the difference between *mass* and *weight*.

Protons and Neutrons

ECOT stresses the fact that the important ID number for an atom or element is the *atomic number*, **Z**. This number means several different things:

- It denotes the position of the element in the Periodic Table in the front of this *Guide*. The smaller of the two numbers in the box surrounding the symbol for the element is the atomic number, Z; the larger of the two numbers is the atomic weight. This Periodic Table will be taken up in greater detail in Chapter Three, *Chemical Bonding*.

- It indicates the number of protons in the nucleus of the atom.

- It represents the number of electrons outside the nucleus of a neutral atom;

- It determines the chemical behavior of that element. It does not matter what the mass number of the element is. If you want to know how an element is going to react chemically, all you have to know is its atomic number.

This last point is easy to understand; chemists realized that the chemical behavior of an element is primarily determined by the number of the electrons in its outermost shell. There is a maximum of 2 electrons in the first shell, 8 in the second, and so on for all the elements in the periodic table as shown in Figure 2.12 of ECOT. The number of electrons in the outside shell determines chemical behavior. Thus the atomic number, Z, which specifies the number of protons inside the nucleus and the number of electrons outside the nucleus, is the determining factor with regard to the chemistry of the element.

Once you understand that Z is the determining factor in chemical behavior, it is easy to accept that a variation in the number of neutrons in the nucleus does not change the chemical identity of the element. ECOT mentions that there are three different "kinds" of hydrogen—protium (with no neutrons), deuterium (with one neutron) and tritium (with two neutrons)—but they all behave the same chemically.

Another example is boron. Whether a boron atom has a mass of 10 or a mass of 11 is irrelevant to the chemistry of boron because all the boron atoms have 5 protons (Z = 5) in the nucleus. Also, the fact that there can be a variable number of neutrons in the nucleus explains why there are fractional mass numbers in the Periodic Table. Some of the boron atoms on this planet have five neutrons in the nucleus and some of them have six neutrons; the mixture of the two kinds of boron is such that the average atomic mass of the mixture of the two is 10.81, listed in the periodic table inside the front cover of this *Guide*.

Newspapers usually print isotopes using symbols such as U-235 or B-11, for instance. This is unambiguous; however, other methods can give more information. Although there is no internationally agreed upon convention, ECOT shows the most often used method of display among nuclear scientists when it shows a form like the one below ("X" is any atomic symbol):

$$\frac{\text{mass number}}{\text{atomic number}}X \qquad {}^{A}_{Z}X \qquad {}^{11}_{5}B \qquad {}^{238}_{92}U$$

The next to last symbol shows that we are talking about an isotope that has 5 protons and (11-5 =) 6 neutrons, named boron, B, and the last symbol designates an atom of uranium which has 92 protons (*all* atoms of uranium, after all, have 92 protons) and a mass number of 238. This means that it has 238 - 92 or 146 neutrons.

Questions Answered

2.2 An atom of gold has a mass of 3.27×10^{-22} g. The ratio can be written

$$\left(\frac{1 \text{ atom}}{3.27 \times 10^{-22} \text{ g}}\right) \text{ or} \left(\frac{3.27 \times 10^{-22} \text{ g}}{1 \text{ atom}}\right).$$

Choose the ratio which, when multiplied by 197 g, will give an answer in "atoms".

$$\left(\frac{1 \text{ atom}}{3.27 \times 10^{-22} \text{ g}}\right) 197 \text{ g} = 6.02 \times 10^{23} \text{ atoms}$$

2.3 a. You are weightless, so the screwdrivers are also weightless. (So is your spaceship, for that matter.)

b. The 100 g screwdriver would need twice as much force to get it up to the same speed as the 50 g screwdriver. (Imagine trying to move your spaceship by pushing on it!)

2.4 a. The mass of the hydrogen atom is equal to the mass of a single proton, or 1 amu.

b. 9 amu + 10 amu = 19 amu.

2.5 Lithium has 3 protons, 4 neutrons, and 3 electrons; ^7_3Li

2.6 No. All H atoms must have 1 proton. This atom is helium, ^3_2He, since it has two protons.

2.7 Eleven electrons must be distributed as follows: 2 (maximum) in the first shell, 8 (maximum) in the second, and since there is one more electron left, the third shell has one.

2.8 a. $100 \text{ paper clips} \times \dfrac{0.100 \text{ g}}{1 \text{ paper clip}} = 10.0 \text{ g}$

b. $\dfrac{\text{total mass of atoms}}{\text{number of items}} = \text{average weight of item}$

c. $\dfrac{10.0 \text{ g}}{100 \text{ paper clips}} = 0.100 \text{ g per paper clip}$

d. $\left(100 \text{ paper clips} \times \dfrac{0.100 \text{ g}}{1 \text{ paper clip}}\right) + 0.200 \text{ g} = 10.2 \text{ g}$

e. $\dfrac{10.2 \text{ g}}{101 \text{ paper clips}} = 0.101 \text{ g per paper clip}$

Supplementary Exercises

S1. An astronaut, undergoing a routine medical examination on earth, is found to weigh 165 pounds (75.0 kg). Given the fact that the force of gravity on the surface of the moon is 1/6 of the earth's force, how much would the astronaut weigh at the moon's Tranquility Base?

S2. The astronaut at Tranquility Base found an interesting piece of rock weighing 5.7 kg. If this rock were transferred to the planet Saturn, where the force of gravity is 1.32 times that of Earth, how much would the rock weigh?

S3. How many protons, neutrons, and electrons are there in each of the following atoms?
a. ^{12}C; b. $^{235}_{92}U$; c. C-14; d. $^{14}_{7}N$.

S4. Using data from Exercise 22 of ECOT, calculate the percentage of oxygen in water. Check your answer by subtracting the percentage of hydrogen in water from 100.

S5. At room temperature, mercury, Hg, is a liquid. Using 2.9×10^{-8} cm for the diameter of a mercury atom and 3.42×10^{-22} g as the weight of one Hg atom, calculate the weight of a 1 cm cube filled with Hg. How does your answer compare with the carefully measured density of 13.59 g/cm^3 for this liquid metal?

S6. Naturally occurring Li is a mixture of Li-6 and Li-7 and the average atomic weight is 6.94. Does the mixture contain more of the Li-6 or Li-7?

Where you might goof up...

- Forgetting the definition of percentage.

- Confusing number of atoms in a formula with weights of atoms in a compound.

- Thinking that a quantum shell is exactly like the earth's orbit around the sun.

- Not being able to give the name, symbol, and the electronic structure for the first 20 elements.

- Confusing the numbers associated with the isotopic symbols.

- Confusing mass and weight.

CHAPTER 3

Chemical Bonding

IONIC AND COVALENT COMPOUNDS

If you had never had a science course and someone asked you what matter is and how it holds itself together, how would you figure it out? Let's try making a model of the problem.

Find an empty shoe box, and ask a friend to put something in the box and tape it shut before giving it back to you. Your job is to figure out what's in the box without opening it. Can you do it? How? What experiments could you perform on the box, short of opening it, that might give you some indication of what's in the box or at least of some properties of the contents? Try it. Some objects are fairly easy. A rolling sound might indicate a marble; something solid that doesn't roll could be a rock with flat sides. But what if your friend carefully cut a block of Styrofoam and fitted it exactly in the box, so that when you shake the box you hear nothing, and it feels like it weighs no more than an empty shoe box? What's your next step?

No doubt you are getting a feel for the difficulty of discerning atomic structure and, more importantly for this chapter, how atoms attach to other atoms. This has been a major question addressed by philosophers and scientists since the 16th century. What accounts for the chemical attraction of one atom for another? On a larger scale, what holds the Universe together?

Seventeenth century philosopher-chemists spent much time and effort arguing the existence of a fundamental particle. Once one decided that a fundamental particle (which couldn't be seen) existed, the next step was to imagine what it looked like. And since the macroscopic world was an obvious (and perhaps the only) pattern for existence, it became the example for the microscopic world. A common idea was that atoms were spheres and that atoms of different substances might differ in color or some other aspect. And then, mirroring the way things work in the macroscopic world, one idea brought forth was that some atoms had hooks and some had eyes. So an atom imagined to look like this, ●—○, might connect with one shaped like this, ⌐○.

But as time passed and science moved away from philosophy and came into its own, experimentalists like Lavoisier, Dalton, Gay-Lussac, Avogadro and many others gave impetus to a new chemistry that, by the middle of the 19th century, was beginning to understand bonding in terms of Faraday's electrical interactions.

Centuries of old ideas have pointed the way to our modern theory of bonding which is based on sound scientific investigation made with the help of modern technology. But remember what we said about theories in Chapter 1 of this Guide. *Theories invite challenge. What will our current theory of bonding look like to a 22nd century chemist?*

Chapter Overview

- An experiment is described which clearly shows that elements do not retain their chemical and physical properties when they become compounds.

- Lewis structures (electron dot structures) are introduced and the periodicity of chemical properties is explored.

- Mendeleev, the periodic table and the idea of chemical families is introduced and expanded upon.

- Valence shells are discussed.

- Building on the idea of electronic structure from Chapter 2 of ECOT, "Atoms and Elements," an explanation of why one Na atom combines with one Cl atom is given.

15

- Valence is related to chemical formulas.

- Chemical formulas are translated into English and back again into formulas.

- A method for predicting chemical formulas for simple compounds is given.

- The difference between ionic and covalent bonds is pointed out.

- Electronegativity and polar covalency are explained, and ionization of polar covalent compounds in water is explored.

For Emphasis

The Periodic Table

When a metal reacts with a nonmetal, an ionic compound results. You might ask, "How do you tell a nonmetal from a metal?" It's easy. Look at the periodic table on the inside front cover of this *Guide*. Then look at the right side of the periodic table in the figure to the right. Using a pen, draw the same stair step line on the periodic table in the front of the *Guide*.

All elements to the left of that line are metals; all those to the right are nonmetals. If a metal on the far left side unites chemically with an element to the far right of the line, the bond between them will be *ionic*. Thus we can predict that compounds such as CaO, $SrCl_2$, RbI, Li_2S, and CsBr will be ionic. Confirm this by finding each of these elements in the table in the front of the *Guide*.

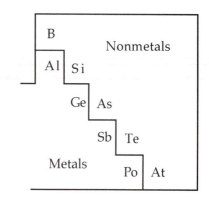

Once you begin to know how to read the periodic table, it will be easy to see how important it is to chemists, and how much we depend on it. The amount of material packed into that one table is simply enormous. As you proceed through your chemistry course you will be told to refer to the table countless times. It has so much information in it that you will find it a great help on assignments and tests.

The atomic number, Z, determines the place of an element in the periodic table. The order of filling the boxes in the table changes one integer at a time. The atomic weight, however, does not always change in an analogous manner. One of the problems with Mendeleev's proposal was that elements that fit according to Mendeleev's idea (chemical behavior), did not fit according to increasing atomic weight. In some cases, the atomic weights had not been well measured; as more careful determinations were made, it was found that the earlier data were not sufficiently precise and the new data fitted properly into the table on the basis of chemical behavior rather than weight. In other cases, some elements are still "out of order" with respect to atomic weight. Co, for example (at. wt. = 58.9), comes before Ni (at. wt. = 58.7). There are several other pairs like this in the table.

Isotopes and Isotopic Abundance

Supplementary Exercise S6 in Chapter Two of this *Guide* discusses the two isotopes of Li, 6 and 7; it asks, "If the average atomic weight of the mixture is 6.94, which is more abundant, ^6Li or ^7Li?" In "Solutions to Supplementary Exercises" in the Appendix, you can see how to reason the correct answer to this question. The next question to be asked, of course, is, "What is the fraction of each in the mixture?" This takes a little algebra, and the steps are very straightforward. Note that the solution below is a more general method of solving a problem like the Example in Section 3.6 of ECOT .

Solution: Since there are two components, the fraction with mass 6 plus the fraction of mass 7 must = 1. (Are you convinced? If so, go on to the next step. If not, think about this: if the fraction of one component is 1/4, the other component has to be 3/4 of the whole; and 1/4 + 3/4 = 1.)

Let y = fraction of mass 6 in the mixture.

Then, $1 - y$ = fraction of mass 7 in the mixture.

Each isotope contributes to the *weighted average* of the naturally occurring mixture by its *weight* and the *fraction* in which it is present. Thus, if ^6Li is present as fraction y, then its contribution to the mixture will be $6y$.

6(fraction of 6) + 7(fraction of 7) = *weighted average* or average at. wt. in amu
Using the numbers:
$$6y + 7(1 - y) = 6.94$$
$$y = 0.06$$

Since we let y = the fraction of mass 6 in the mixture, multiplying 0.06 by 100 gives us the percentage of ^6Li in the mixture, or 6%; ^7Li is thus 94%. This works for any other naturally occurring isotopic mixtures having two components.

If you have done the problem correctly, your answer must be consistent with the information given in the problem. Check it using the same kind of thinking that we did in the answer to S6 in the Appendix.

The change in chemical behavior between some elements and their compounds is one of the remarkable things about the study of chemistry. Sulfur dioxide is extremely corrosive to lung tissue and is a severe eye irritant, but elemental sulfur when made into a salve is used for certain skin ailments. Comparisons can be made between elements and some of their compounds by browsing through the *Merck Index*.

Lewis Structures (Electron Dot Notation)

Table 3.1 in ECOT provides Lewis structures for the first 20 elements. You should practice writing these and, from an element's position in the periodic table, be able to tell how many electrons it has in its valence shell. In other words, you should be able to use the *periodicity* of the elements to predict the Lewis structure of any element in the families of elements discussed in ECOT. For example, if you know that magnesium, Mg, has 2 valence electrons, and you see in Table 3.1 that calcium, Ca, does also, you know from the *periodicity* of chemical properties that strontium, Sr, barium, Ba, and radium, Ra, should also have 2 valence electrons, and their chemistry should be similar to that of Mg.

Writing Chemical Formulas

Writing the formulas of binary compounds is easy, and you will soon be writing them without going through the arithmetic on paper. As the author of ECOT says, "charges must equal" so let's see how we can make this simple.

Let's review what we know about some of the elements. We know from the Lewis structure of oxygen that it has 6 valence electrons. In combination with other elements oxygen generally gains two electrons to form the oxide anion, O^{2-}. The halogens generally gain one electron to form F^-, Cl^-, Br^-, I^-, At^-. Since the entire halogen family reacts similarly, what would you expect to be true of the family containing oxygen? Since oxygen gains two electrons, sulfur, S, probably does also, as should Se, Te, and Po. (See how useful the periodic table is?) To find out for certain that all the elements in the oxygen family do as we predict, we would have to do some experimentation. (Remember, chemistry is an *experimental* science!)

ECOT's formula determination method, Section 3.12, is very simple. The only thing you have to remember is that for a compound the total positive charge must equal total negative charge. So if the ions are Ca^{2+} and Cl^-, it will take 2 Cl^- ions for each Ca^{2+} ion, and the formula, writing the cation first, is $CaCl_2$.

What if the charges are not given? Let's write the formula for calcium oxide.

> **Solution:** From the periodic table, you know that calcium, Ca, has two valence electrons and can transfer those to oxygen. An oxygen atom has six valence electrons and can accept the two electrons from Ca. So the formula will be CaO.

Now write the formula for magnesium iodide.

> **Solution:** From the periodic table, you know that magnesium, Mg, has two valence electrons and can transfer those to iodine. But an iodine atom, with seven valence electrons, can accept only one electron from Mg. So it will take two iodines to accept the two electrons from a magnesium atom. The formula is MgI_2.

What about sodium oxide?

> **Solution:** Sodium has one valence electron and can transfer that one to oxygen. But an oxygen atom, with six valence electrons, needs two electrons to obtain an octet. So it will take two sodiums for each oxide. The formula is Na_2O.

How do we find the valence of an element in a compound? To do this you will always have to know the valence of one of the elements in the compound. For example, what is the valence of manganese, Mn, in manganese sulfide, MnS?

> **Solution:** Manganese is a metal, but it is not in the alkali metals or the alkaline earths. However, we know that sulfide is analogous to oxide, so it will be S^{2-}. In this compound, then, Mn must have a charge of 2+.

Questions Answered

3.1 Look for the endings *-ine* or *-ide*. Iod*ine* is the element, iod*ide* is the ion. Thus, a. antiseptic; b. potassium iodide; c. both antiseptic and potassium iodide (note that in KI we have both the element, I, in the form of the iodide ion, I^-).

3.2 Sodium chloride, table salt, NaCl.

3.3 From the elements and compounds mentioned so far in ECOT you might obtain the following answers. There are others, of course, so try to think of them.

An element harmless to us in its elemental form: nitrogen
An element hazardous in its elemental form: chlorine
A hazardous compound of the harmless element: nitrogen dioxide
A harmless compound of the hazardous element: sodium chloride

If you wish to go beyond the ECOT text, *The Merck Index* is probably in your library, and it is a good source of information that you may find useful beyond this course. The technique for using *Merck* is: a) look up an element and read toward the end of the entry where cautionary advice is printed; then b) peruse following pages for entries of compounds containing that element which provide poison and toxicity information. For instance, *carbon*, entry #1814 in the *Eleventh edition* is described as an antidote for ingested poisons whereas entry #1820, *carbon monoxide*, has an entry mentioning death.

3.4   H, Li, Na, K; F, Cl; Ne, Ar

3.5   Be, Ca. They are in the same family as magnesium, Mg.

3.6   Li-6 is probably the other isotope. If the contribution of Li-7 is 92.5%, the contribution Li-6 is 7.5% of the mixture. Then the contribution of both of them to the naturally occurring mixture will be:

$$\text{contribution from Li-6} = 0.075 \times 6 \text{ amu} = 0.45$$
$$\text{contribution from Li-7} = 0.925 \times 7 \text{ amu} = \underline{6.48}$$
$$\text{sum of both masses} = 6.93$$

This is close to the reported average atomic weight, 6.94 amu.

3.7   Beryllium, Be.

3.8   a. 2;   b. 2;   c. Yes; they are expected to react in a one-to-one ratio. Since one is a metal and the other is a non-metal, an ionic compound should form as Mg donates 2 electrons to O. (Its formula will be MgO.)

3.9   Neon; argon.

3.10   Snowflakes.

3.11   a. Magnesium iodide; b. beryllium sulfide; c. potassium oxide. Urea: CH_4N_2O. Look at rule #2 for writing formulas for compounds containing carbon: C first, H second, all others in alphabetical order.

3.12   As in the examples above, keep in mind that the total charge on any molecule is zero.

a. Each oxygen has a charge of 2- giving a total of -4 for the oxygen (the "2-" is the symbol dictated by convention for a charged species, the "+4" is a quantity of charge); to balance this, the Pt must have a charge of +4, labeled 4+;
b. Since oxygen has a charge of 2-, each Cu has a charge of 1+;
c. Sulfur is in the same family as oxygen, so it has a charge of 2-; to balance this, Hg must have a charge of 2+.

3.13   a. 10; b. 20; c. C_3H_8

3.14   b, d, f.

3.15   HBr; 0.1 g HF produces only a few H^+ and F^- ions when dissolved in water, but 0.1 g HBr completely ionizes to produce many H^+ and Br^-. Water will not conduct an electric current without ions and when more ions are present, an electric current is conducted more effectively.

3.16   Ferric oxide has the formula Fe_2O_3. Since oxygen will have a -2 charge, and there are 3 oxygens, the total negative charge in the formula is -6. The total positive charge must be +6, so the charge of each of the two iron ions must be 3+, or a valence of +3.

Supplementary Exercises

S1. Two isotopes of B, 10 and 11, are found in naturally occurring boron, which has atomic weight 10.81. What is the isotopic distribution in this mixture?

S2. Write the equation twice for the reaction between K and F to give potassium fluoride, once with the elemental symbols (don't forget charges) and once using Lewis dots.

S3. Which of the following compounds when mixed in water forms a solution that conducts electricity? a. KBr; b. CH_4N_2O; c. CH_4; d. HBr; e. CCl_4; f. RaI_2.

S4. Write the names for the following formulas: a. Li_2O; b. $CH_3CH_2CH_3$; c. Al_2S_3; d. H_2S; e. $KMnO_4$; f. NS_2.

S5. Using lines for covalent bonds, sketch the structural formula for C_2H_6. How many covalent bonds does this molecule have? How many bonding pairs of electrons? How many total electrons?

S6. What is the valence of each cation in the following compounds? a. BaF_2; b. VCl_5; c. OsO_4; d. WO.

S7. What is the valence of the anions in the following compounds? a. Na_2Se; b. $MgBr_2$; c. Mg_3N_2; d. Ca_2C.

S8. One of the two isotopes in naturally occurring chlorine, Cl-35, comprises 75.8% of the mixture which has an average atomic weight of 35.45. What is the mass of the other isotope?

S9. Calculate the average atomic weight of rubidium if the naturally occurring mixture is comprised of 72.2% Rb-85 and 27.8% Rb-87.

Where You Might Goof Up...

- Not being able to describe in detail the demonstration described at the beginning of the chapter in ECOT because you did not take the time to do it yourself.

- Failing to distinguish between ionic and covalent bonds.

- Being unable to distinguish metals from nonmetals in the periodic table.

- Not practicing Lewis dot notations for the first 20 elements and some of their binary compounds.

- Being unable to use the word "isotope" correctly or to calculate the percentage of isotopes in a mixture.

- Not being able to identify a family in the periodic table, and failing to explain adequately why family members behave the same chemically.

- Being unable to use the periodic table to help you determine valence.

CHAPTER 4

Discovering the Secrets
of the Nucleus

FROM A PHOTOGRAPHIC
MYSTERY TO THE ATOMIC BOMB

In the 1920's it was fashionable for the wealthy to carry pocket watches rather than wristwatches, which were just becoming popular. It was even more fashionable to have a pocket watch with a hand-painted dial; it was especially fashionable to have a pocket watch with a hand-painted dial that glowed in the dark. To make the numbers on a watch dial glow in the dark a mixture of radioactive radium and zinc sulfide, ZnS, was incorporated in the paint base; when the radium gave off its energy, it excited the ZnS, which then glowed. To satisfy the demand for this product, whole factories were populated by young women painting dials with the radioactive mixture. At the time little was known of the hazard of handling radioactive substances, and the Occupational Safety and Health Administration (OSHA) wouldn't be a reality for another forty years or so. In order to do the most delicate work, many of the women twirled the paint brush hairs on the ends of their tongues to make a fine point, thus ingesting small quantities of radium. After a while, their bone marrow was damaged and many died of leukemia.

Chapter Overview

• Becquerel's serendipitous discovery is described, as are the Curies' experiments.

• Alpha (α), beta (β), and gamma(γ) radiation are introduced as well as some of their properties, particularly their penetrating properties.

• The method that nuclear physicists and radiochemists use to write chemical symbols is shown and explained with a few examples.

• Balancing nuclear equations is demonstrated.

• Transforming one element into another (with a change in chemical properties, naturally) is discussed.

• Quarks and gluons, particles that are thought to make up the tiny proton and neutron, are introduced.

• The Manhattan Project, the building of the bomb, and its delivery are discussed.

• The equivalence of mass and energy is noted and calculations relating to the disappearance of one and the appearance of the other are shown.

• The hydrogen bomb, an even more powerful "device," is described.

For Emphasis

Types of Radiation

Once you have read and "quizzed the chapter" it's important to go back to the properties of the various types of ionizing radiation. These are referred to many times in this chapter and in Chapter 5, "Energy, Medicine, and a Nuclear Calendar," and it is important for you to be fluent in their properties. A convenient way of thinking about and remembering these is by constructing an information table. This is usually a vertical column of items connected to a horizontal list of properties or other important data; Table 4.1. in ECOT is one form of this. But, *without looking at Table 4.1 in ECOT* reread section 4.4, "α Particles, β Particles, and γ Rays" very carefully, and, as you read, fill in the holes in the table below, and make any comments about the radiation that are not covered by the columns of the table. The result will look very much like Table 4.1, but you will have done it from the reading; thus, you will have learned it as you picked out the information.

Type	Mass	Charge	Penetrating Power	Symbols	Velocity	Composed of...	Comments
	0		moderate				
	0				speed of light		
α						He nucleus	

Chemical and Nuclear Reactions

To track what is occurring in nuclear reactions we use nuclear equations to tell us what we start with and what we end up with. The word "equation" implies that there is an equality, and indeed there is. Like the simple chemical equations you saw in Chapter 3 (the ionization of HCl, for example), nuclear equations are written with what we start with on the left, an arrow to indicate that a reaction has proceeded, and what we got out of the reaction on the right. (In chemical parlance these beginning and ending substances are called *reactants* and *products*, respectively.) The arrow is not an equal sign, but it does imply that something is equal. That "something" is the quantity of matter in the reaction. The Law of Conservation of Mass (see Section 4.11 in ECOT) says that matter can neither be created nor destroyed as a result of *chemical* transformations. Parallel with this is the Law of Conservation of Energy which states that energy can neither be created nor destroyed as a result of *chemical* transformations.

What's the difference between a chemical and a nuclear reaction? In Chapter 3 we were using Lewis structures to describe what electrons do in forming chemical bonds. The key to *chemical* reactions is understanding that they are about electrons. Substances undergo chemical reactions by breaking and forming chemical bonds in which electrons are involved. The nucleus is *not* involved. It is unchanged by the reactions the atom is undergoing as electrons in the quantum shells are gained, lost, or shared.

The transformations we are studying in this chapter are *nuclear*, not chemical, and we find that the Law of Conservation makes a different statement where nuclear reactions are concerned; it becomes the Law of Conservation of Mass *and* Energy. Unlike chemical reactions where matter is not created or destroyed, in nuclear reactions a very small quantity of matter can be converted into energy or vice versa. Now our concern is not with the quantum shells but with the interior of the nucleus itself. In fact, in one sense, we are no longer doing *chemistry*, since chemical reactions do not change the nucleus. But there is a branch of chemistry called *radiochemistry*. Radiochemists or nuclear chemists study the *chemistry* of the products of nuclear reactions. Nuclear physicists are the people who look for the quarks and gluons and the myriad other infinitesimal nuclear parts that are essential for rounding out our understanding of matter.

One could say, then, that in studying the nucleus and radioactivity in this chapter and the chemical and biological effects of radioactivity on living tissue in Chapter 5, we are right at the interface of three great disciplines: biology, chemistry, and physics. And what an exciting interface it is! Not only are we studying the destructive power of the nucleus, but we will also see the nucleus used to diagnose illness and provide treatment for cancer.

Balancing Nuclear Equations

So, let's get back to the subject of nuclear reactions so that all those interesting things to come will make sense. The key to writing and balancing nuclear reactions is in knowing the information in the table you made on the previous page. Let's look at a simple nuclear word equation.

> Tritium decays by beta emission to form an isotope of helium which contains only one neutron.

Now to write it in symbols:

$$^{3}_{1}\text{H} \ \rightarrow \ ^{3}_{2}\text{He} + ^{0}_{-1}\beta$$

or

$$^{3}_{1}\text{H} \ \rightarrow \ ^{3}_{2}\text{He} + ^{0}_{-1}\text{e}$$

There are some important features to understand. Note that mass is conserved; that is, we have a mass number of 3 on the left and a total mass of 3 + 0 or 3 on the right. What happens to the atomic number or total positive charge? Look at the lower numbers. 1 = 2 + (-1). Where does the beta particle, an electron, come from? Beta emission requires that a neutron be converted into a proton + an electron. The electron is emitted from the nucleus, and the nucleus now has an additional proton, so the atomic number increases by 1, which means we have a new element. So, a rule :

> Rule #1: When beta, β, emission occurs *the mass number stays the same* and *the atomic number increases by 1*. Look on the periodic table to identify the new element.

What about alpha emission? It's often useful to get in the habit of using the element symbol notation for the alpha particle rather than the Greek letter, α, in the equation.

$$^{238}_{92}\text{U} \ \rightarrow \ ^{234}_{90}\text{Th} + ^{4}_{2}\text{He}$$

Writing the a particle as the helium nucleus that it is allows us to do the accounting we need to do to determine that mass has been conserved and that the atomic number or positive charge is consistent. You can see that 238 = 234 + 4, and that 92 = 90 + 2. Note that the new element formed, thorium, has atomic number of 2 less than uranium.

> **Rule #2:** When alpha, α, emission occurs *the mass number decreases by 4* and *the atomic number decreases by 2*. Look on the periodic table to identify the new element.

Gamma emission accompanies most radioactive decays. Because it is energy and not matter, its emission does not change the atomic number or mass number of the species involved. (Look at your table to confirm that it has no mass or charge.) Therefore, it is generally not shown in the reactions above. It becomes more important in the fusion reaction of section 4.14.

Questions Answered

4.1 *Incorrect assumption*: Phosphorescing substances emit X-rays.
Correct conclusion: The uranium compound caused a developed spot on a photographic plate.
Incorrect conclusion: The uranium compound emitted X-rays while it glowed (phosphoresced) after being exposed to sunlight.
Key result: The uranium compound developed a photographic plate without being exposed to sunlight.

4.2 α rays; α rays and β rays; γ rays

4.3 6, 7, 8

4.4 Atomic number: a. decreases by 2; b. increases by 1; c. stays the same;
mass numbers: a. decreases by 4; b. stays the same; stays the same.

4.5 $^{210}_{84}Po \rightarrow ^{206}_{82}Pb + ?$ To decide what the particle is, look at the atomic number and the mass number. On the right side of the arrow we need Z = 2 and A = 4, so this must be an α particle. The equation, then, is $^{210}_{84}Po \rightarrow ^{206}_{82}Pb + ^{4}_{2}He$.

4.6 Write the equation as in the problem above. In this case Z = -1 and A = 0, so it must be a β particle.

4.7 The nucleus of hydrogen is a positively charged proton, and it's surrounded by a negatively charged electron. The symbols are $^{1}_{1}p$ and $^{0}_{-1}e$.

The nucleus of antihydrogen is negatively charged (antiproton instead of a proton), and it's surrounded by a positively charged region (positron instead of an electron). Its particle symbols are $^{1}_{-1}p$ and $^{0}_{1}e$.

Antihydrogen and hydrogen (protium) both have a mass number of 1 because an antiproton is identical to a proton except for the charge.

4.8 The atomic numbers of the products (remember that the "atomic number" of a neutron is zero) must add up to the atomic number of the fissile nucleus. Thus, given that Z for Ba is 56, 92 - 56 = 36 which is krypton.

4.9   a. The advantage of using U-235 is that fission of a U-235 nucleus releases neutrons that cause fission in other nuclei, thus starting a chain reaction.
b. The disadvantage of using U-235 is that it is in such low concentration in the naturally occurring mixture (0.7%). It must be enriched to at least 80% to make even a marginally efficient bomb.

4.10 In the U-235 bomb (Little Boy), a subcritical uranium slug was fired into a subcritical ring of uranium to form a critical mass of U-235. In the Pu-239 bomb (Fat Man), a hollow sphere of plutonium was explosively crushed into a pellet that had the correct mass to cause a chain reaction. (Look carefully at Figure 4.11.)

4.11 In a chemical reaction, the mass of the reactants must equal the mass of the products. Similarly the total energy of the reactants in a chemical reaction must equal the total energy of the products. In nuclear fission, however, some mass is converted into energy, so the mass of the starting materials is greater than the mass of the products. This finding directly violates the Law of Conservation of Mass. Similarly, the energy of the reactants is less than the energy of the products (some mass is converted into energy) and this violates the Law of Conservation of Energy. Therefore, fission cannot be a chemical reaction.

4.12

94 protons x 1.007=	94.658 amu
145 neutrons x 1.009=	146.305 amu
94 electrons x 0.0005=	0.047 amu
calculated mass (sum) =	241.01 amu
measured mass =	239.05 amu
mass defect (difference)=	1.96 amu

4.13 Table 4.2 shows that converting 1g of matter (*any* matter) into energy would keep the bulb burning 29,000 years.

4.14 Total protons on the left is 3, on the right 1, so Z of the particle =2; Total mass on the left is 7, on the right is 3, so 4 amu are needed. Thus the particle is an alpha particle. Write the equation for practice.

Supplementary Exercises

S1.   List the number of protons, neutrons and electrons in each of the following, which occur naturally: a. U-238;   b. F^-;   c. $^6Li^{3+}$, d. ^{132}Xe.

S2.   Fill in the blank space in each of the following equations (in *e* place the Greek letter of the particle emitted over the arrow).

a. $^{20}_{9}\text{F} \longrightarrow + \beta$

b. $^{168}_{70}\text{Yb} \longrightarrow + \gamma$

c. $^{2}_{1}\text{H} + \longrightarrow \, ^{1}_{1+}\text{p} + \, ^{14}_{7}\text{N}$

d. $^{206}_{83}\text{Bi} \longrightarrow \, ^{206}_{82}\text{Pb}$

e. $^{210}_{83}\text{Bi} \longrightarrow \, ^{210}_{84}\text{Po}$

S3.   Given the mass of Br-89 to be 88.89 and La-144 to be 143.901, calculate the loss in mass for the following fission reaction: $\text{n} + \, ^{235}\text{U} \rightarrow \, ^{144}\text{La} + \, ^{89}\text{Br} + 3 \, \text{n}$

S4.   Below is the first part of the U-238 radioactive decay series shown in Figure 4.7 in ECOT. Over the first two arrows identify the type of emission (α or β) and fill in the blank between the last two arrows with the correct isotope.

$$^{238}_{92}\text{U} \longrightarrow \, ^{234}_{90}\text{Th} \longrightarrow \, ^{234}_{91}\text{Pa} \xrightarrow{\beta} \xrightarrow{\alpha} \, ^{230}_{90}\text{Th}$$

S5.   Figure 4.7 shows a radioactive decay chain in which decay begins with the emission of an alpha particle from U-238 and ends with the stable isotope Pb-206. A similar sequence of alpha and beta emissions occurs in a decay chain which starts with the thorium isotope, Th-232. The isotope at the end is another stable isotope of lead, Pb-208. Following is the Th-232 decay chain with parts omitted. Complete the series with either the type of emission (alpha or beta) over the arrow or the isotope produced in the space provided.

$$^{232}_{90}\text{Th} \xrightarrow{\alpha} \, ^{228}_{88}\text{Ra} \longrightarrow \, ^{228}_{89}\text{Ac} \xrightarrow{\beta} \longrightarrow \, ^{224}_{88}\text{Ra} \xrightarrow{\alpha}$$

$$^{220}_{86}\text{Rn} \longrightarrow \, ^{216}_{84} \longrightarrow \, ^{212}_{82}\text{Pb} \longrightarrow \, ^{212}_{83}\text{Bi} \xrightarrow{\beta} \longrightarrow \, ^{208}_{82}\text{Pb}$$

S6.   The introductory paragraph in this chapter described a tragic affair of radiation poisoning. Explain how ingested radium could cause leukemia and death. (Hint: start by locating Ra in the periodic table and identifying a biologically important element in the same family.)

Where You Might Goof Up...

- Forgetting that energy and matter are interchangeable.

- Having a poor command of nuclear notation.

- Not bearing in mind that if a charged particle radiates from the nucleus, Z changes and so does the chemical nature of the new atom.

- Not balancing both mass number and atomic number in nuclear equations.

- Not being able to distinguish between *fission* and *fusion*.

- Not being able to convince your instructor with words and diagrams that you understand what a chain reaction is.

- Not knowing which isotope moves faster in a gaseous diffusion process.

- Being unable to explain *qualitatively* and *quantitatively* where nuclear energy comes from.

CHAPTER 5

Harnessing the Secrets of the Nucleus

NUCLEAR ENERGY, NUCLEAR MEDICINE, AND A NUCLEAR CALENDAR

Once upon a time to come:

- *5 to 10 seconds: Core temperature rises to 850 °C from a normal 360 °C in certain sectors of the reactor vessel.*

- *0 to 60 seconds: Temperature rises above 1150 °C and core structure is attacked by steam reaction; energy release rate rises.*

- *50 to 100 seconds: Core temperature rises to 1900 °C.*

- *2 minutes: Core collapse commences.*

- *10 minutes: Meltdown debris accumulates in bottom of reactor vessel.*

- *1 hour: Melt-through of pressure vessel is anticipated with attendant steam explosions.*

- *1 day: Molten mass of reactor material burns through outer containment slab (China Syndrome).*

- *3 years: Molten innards of reactor vessel form 36-meter diameter blob in earth.*

- *11 years: Blob cools somewhat and shrinks to 29 meters.*

The above is taken from a scenario imagined over thirty years ago by Ralph Lapp, a nuclear scientist present at the outset of the nuclear age. It is a tribute to his understanding of the science and engineering of nuclear power stations that the limited but practical experience we have had since then with "excursions," as the nuclear power community calls them, has resulted in only a small change in one of the temperatures mentioned above. The question is, as a member of the power station control room team, "What is the latest you can intervene in the time line above and avert disaster?"

Chapter Overview

- Details of the first controlled nuclear reaction are given.

- It is pointed out that a nuclear power station is just another steam plant with different fuel.

- The promise, costs, and problems of using nuclear fuel are described.

- The downside of nuclear power plants as was shown at Ottawa, Three Mile Island, and Chernobyl is recounted.

- The breeder reactor as a source of nuclear fuel is described.

- The problem of nuclear wastes is presented.

- The upsides of nuclear power and nuclear medicine are related.

- Half-life of a radioisotope is introduced and explained.

- Methods for detecting radiation are described.

- Radioisotope dating is detailed.

For Emphasis

Half-Life: $t_{\frac{1}{2}}$

You will understand the concept of half-life if you keep in mind that it is a unit of *time*, not a quantity of material or a rate of ejection of radiation. Figure 5.2 in ECOT shows the data in Table 5.4 of ECOT plotted for the first six half-lives. Each half-life which passes (8 days, in this case) means that half of the radioactive iodine has decayed into something else. Implicit in this analysis is the fact that the *radiation* coming from the sample is also *halved*. So we could have just as easily labeled the vertical axis "activity", "counts/min," or "disintegrations/hour".

We can use the graph in Figure 5.2 as a "generic" half-life graph in which the vertical axis is relabeled "% of sample remaining." The horizontal axis remains the same. We can convert half-life into seconds, minutes, hours, days, years, whatever is required by a specific problem. Note that the symbol for half-life in arithmetic equations is $t_{\frac{1}{2}}$.

Let's say we placed an unknown radioactive sample in a Geiger counter that initially registered the activity of the sample at 2300 counts/min. The counter registered 950 counts/minute after 35 minutes.

   a. How much of the radioactive sample is left?
   b. How many half-lives have passed?
   c. What is the value of $t_{\frac{1}{2}}$ for this sample?

Solution:

a. $\%$ material left $= \left(\dfrac{950 \text{ cts/min}}{2300 \text{ cts /min}} \right) 100\% = 41\%$

b. Find 41% on the vertical axis. Move horizontally until you hit the curve, and then, at that point, drop down to the horizontal axis. Approximately $1.3\,t_{\frac{1}{2}}$ has expired.

c. Ask yourself a question: If 35 min represents 1.3 half-lives, then how many minutes represent 1 half-life? One way to answer this is to set up a proportion:

$$\left(\frac{35 \text{ min}}{1.3 \text{ } t_{\frac{1}{2}}}\right) = \left(\frac{x \text{ min}}{1 \text{ } t_{\frac{1}{2}}}\right)$$

$$x = 27 \text{ min}$$

This is a legitimate way of thinking, because the stream of radiation (counts/min) is proportional to the quantity of radioactive material present in the sample at any time.

Decay Series and the Rate Determining Step

One of the data in the calculation of the age of the earth involves the ratio of ^{238}U to ^{206}Pb (Section 5.10 of ECOT). From this ratio we can determine an approximation of the age of the earth, since it is assumed that the ^{238}U was formed along with this planet and the *only* source of ^{206}Pb is from the decay of the originally formed ^{238}U. Figure 4.7 of ECOT shows the several intermediate nuclides in this decay scheme, all being radioactive except the last one. It appears to ignore the fact that the intermediate nuclides formed take time to decay, the only half-life used in the calculation being that of ^{238}U, 4.5×10^9 years. It's easy to see why the 5-day half-life of the beta decay of ^{210}Bi, one of the products of the decay of ^{238}U in Figure 4.7, can be ignored, but why is it unnecessary to include the half-life of, say, ^{234}U, which is 2.5×10^5 years?

Let's look at it this way. Consider five workers in a large cafeteria who have the following jobs: Three clear the tables of dishes and deliver them to one who loads two commercial, efficient automatic dishwashers and one dries the dishes by hand and stacks them on shelves. If we define the job as "finished" when all the dishes are stacked and dried, who is holding up the parade? The dish dryer, of course, who is involved in the slowest step. No matter how fast the other workers pick up and deliver the dishes or load the washer, the dish dryer controls the speed with which the job is finished. The last step in this case is the *rate-determining* step. Now suppose we redistribute the work load as follows: One clears all the tables, two load the dishwashers and two dry and stack. The rate-determining step this time is clearly the first one. This second scenario more closely approximates the model shown in Figure 5.7 of ECOT and the decay scheme for ^{238}U. Since the step from ^{238}U to ^{234}Th, the first one, is the slowest, it is the only one that needs to be considered.

Questions Answered

5.1   a.   According to ECOT, there were 438 nuclear power plants operating in the word, with 104 of them in the United States.

$$\frac{104}{438} \times 100 = 24\%$$

b.   The US ranks 5th in the world in use of nuclear energy for electric power production.

5.2   The accidental release of radioactive material and the subsequent fallout.

5.3   Advantages of a breeder reactor are that it produces more fissionable material than it uses, it has a long operating life and low operating expenses. Disadvantages include high initial costs, the political danger of large-scale production of plutonium that can be made into a nuclear bomb, and the high toxicity of microgram quantities of plutonium.

5.4   Becoming a stable nuclide.

5.5 All of the nuclides in Table 5.3 in ECOT which have half-lives of seconds, minutes, hours, and days would be long past the 10 half-lives point of negligible radiation. Half of the C-14, on the other hand, would have decayed into other materials after 5730 years. Another half would almost be gone in 10,000 years. In other words, over 25% would remain. (For help with this, see Table 5.3 in ECOT.) So C-14 and all the nuclides in Table 5.3 with half-lives *longer* than that of C-14 would remain.

5.6 a. Disposal of radioactive waste; b. air pollution control, cost of fossil fuels

5.7 Genetic damage and somatic damage.

5.8 a. Because it concentrates in the thyroid gland, I-131 is used to examine thyroid function.
b. Tc-99*m* is used to generate diagnostic images.
c. Co-60 is used as radiation therapy to destroy cancer cells by γ radiation.

5.9 Since positron emission converts a proton into a neutron, the atomic number decreases by one, so the element produced when N-13 emits a positron is element #6, C. The nuclear equation is

$$^{13}_{7}N \rightarrow {}^{13}_{6}C + {}^{0}_{1+}e$$

Note that the symbol for a positron is that of a positive beta particle (see Sections 4.7 and 4.14). The positron is, after all, identical to an electron in mass, but opposite in charge.

5.10 Look at the Example "Old Rocks." Since the half-life of U-238 is 4.5×10^9 years, the number of half-lives in 18.0×10^9 years would be

$$\text{Number of half-lives} = \left(\frac{18.0 \times 10^9 \text{ years}}{t_{\frac{1}{2}}} \right) = 4$$

Now look at the table in that Example. The next entry under "Half-lives of U-238" would be 4, which is what we have determined. The number of U-238 atoms remaining must be half of what it was for 3 half-lives, so there are 4 remaining. If 4 U-238 atoms decay, 4 Pb-206 atoms are formed, so there are now 60 atoms of Pb-206. The Pb/U ratio is 60/4 or 15.

5.11 b,c,d. All of these contain materials from living systems that once ingested ^{14}C.

5.12 a. Geiger counter; b. film badge

5.13. a. Originates in human activities: medical, consumer; b. natural: radon; soil, rocks, atmospheric; internal; miscellaneous.
b. Natural sources make up 82% of the total, while human activities produce only 28% of the total exposure.

Supplementary Exercises

S1. Cadmium is used in control rods of nuclear power plants because it has such a large neutron collision cross section. In other words, a Cd nucleus captures neutrons efficiently. Complete the equation below with the product of neutron capture:

$$^{112}_{48}Cd + {}^{1}_{0}n \rightarrow$$

S2. Nuclear engineers are interested in two types of fuel for their plants, *fissile* and *fertile*; the fissile or fissionable materials include U-235 and Pu-239, whereas fertile materials can become fissile by nuclear bombardment followed by other (breeder) reactions in the reactor core or industrial plant. The fissile ^{233}U can be made from fertile ^{232}Th with a two-step process. The first step is: $^{232}_{90}Th + ^{1}_{0}n \rightarrow X$. Identify X.

What is the final step?

S3. A sample of ^{23}Na decays by positron emission to 25% of its original amount in 62 months. Write the equation for this decay and calculate the half life of this nuclide.

S4. Radiochemists have made an isotope of gold, ^{199}Au, by the bombardment of ^{198}Hg with neutrons. Write a balanced nuclear equation for this process. Speculate on the economic viability of this reaction to produce cheap gold.

S5. The two main radioactive species carried in the plume of smoke and particles generated by the accident at Chernobyl were ^{131}I and ^{137}Cs. Both form soluble compounds, and they fell to the ground in the surrounding area and were incorporated into the food chain. In which families of elements intimately associated with the chemistry of life do each of these nuclides belong? If active cleanup were not pursued, how long would the Ukrainians have to wait before moving back in, planting crops, and undertaking other normal activities? (^{137}Cs has a half-life of 30 years.)

S6. When ^{233}U undergoes typical chain reaction fission (spend one neutron to produce three), one of the fission products is ^{160}Sm. Write the balanced nuclear equation for this process and identify the other product nucleus.

S7. At a dig in the remote part of Iran, archeologists have found a parchment which they believe may be the record of a Mithraic sect, an ancient Persian religion, said to be active around 120-195 CE*. Since no confirmed written records of the cult of Mithras exist, this may be an important find. As a radiochemist you are asked to determine the approximate age of the document. You reduce a small sample of the parchment covering to pure carbon for radiocarbon dating. Your nuclear counter, showing a background radiation of 6 counts per minute (when no sample is in the instrument) shows 35 counts per minute with the document wrappings and 42 counts per minute with the same weight of carbon from a freshly harvested source. What will you write in your report ?

S8. Three people fishing from a boat have the following responsibilities: the first catches the fish, the second carefully cleans the fish and the third packs the cleaned fish in ice. Where is the bottleneck when (a) the fish are not biting well, and (b) when they are biting like crazy?

Where You Might Goof Up...

- Writing a serious essay about a nuclear power plant blowing up like an atom bomb.

- Thinking that there is something "magic" about nuclear power rather than its being another source of heat.

- Being unable to list other fissile nuclides than ^{235}U, such as ^{233}U or ^{239}Pu.

- Being unable to list both the up- and down-sides of breeder reactors.

- Not being able to explain how a Geiger counter works.

- Not having the ability to distinguish between ionizing power and penetrating power of the various kinds of radiation.

- Confusing the nature and consequences of somatic *vs.* genetic damage.

- Forgetting that you may be asked to list and discuss three or four applications of nuclear knowledge to medicine.

- Not being able to deal with straightforward half-life problems.

- Being unprepared to pick the rate-determining step from a sequence of events.

- Being ignorant of sources of natural background radiation.

---

* CE: Common Era. This term is slowly replacing the more parochial A. D. "B. C." then becomes BCE, Before the Common Era.

CHAPTER 6

Introduction to Organic Chemistry

THE POWER OF HYDROCARBONS

August Kekulé sat before the fireplace, staring into the dancing flames. He was tired. Since 1856, when he had announced his theory of the tetravalency of carbon, he had struggled to understand how benzene could have the empirical formula CH. Mesmerized by the flames, he fell asleep. In a dream chains of molecules danced before him, twisting and turning like snakes. Suddenly, one of the snakes grasped its own tail in its mouth, forming a ring. Kekulé awoke with a start. Carbon atoms in benzene form a ring!

Kekulé published his ring theory in 1865. At a celebration of his work in 1890 he told his dream story for the first time. The story has produced a great deal of controversy over the years. Did Kekulé make up the whole thing? Or was he really just daydreaming? Can dreams inform our scientific endeavor?

These are all interesting questions, but the real value of Kekulé's dream story is that it shows us a chemist seeking a model. Kekulé knew from the mass ratio that the empirical formula was CH. He wanted to know how to model a molecule with that empirical formula, using his understanding of atoms and bonding, and of the tetravalency of carbon. The benzene ring does just that.

Models help us visualize concepts that are abstract. We build a spacecraft that can leave the Earth only after we have done calculations and drawings and made models of the craft; we model the long-term effects of volcanic eruption using current atmospheric conditions and extrapolating to the future; we use the computer to build our quantum model of the atom from mathematical constructs, giving the calculations physical reality.

Whether our models are sophisticated computer-generated images or simple entities made of gumdrops and toothpicks (see below), they present us a picture of reality that numbers or drawings on a printed page can't possibly duplicate. They help us see what we can't see. Give yourself the gift of visualizing the invisible by making models of the compounds in Chapter 6 of ECOT.

Chapter Overview

- A demonstration is described of burning a candle in such a way that one of the combustion products is easily identified.

- Organic chemistry is defined.

- Bonding and nomenclature of simple organic compounds is described, and free radicals are explored.

- The existence of isomers is revealed and expanded upon.

- IUPAC, the International Union of Pure and Applied Chemistry, proposes systematic rules for naming compounds.

- Examples of balancing equations are given.

- The double bond is introduced with a description of the alkene and alkyne family.

- Two other families, cycloalkanes and aromatics are presented.

- Functional groups are delineated.

- The greenhouse effect is examined.

For Emphasis

Model the Tetrahedron

You can get a three-dimensional picture of the bond angles around carbon by turning to the last page in the Appendix. There you'll find a page that can be torn out of this book, glued to a thin card and cut out along the lines indicated. Follow the instructions on the page with the diagram to assemble the shape. The corners (tips) of this tetragon show the positions of the hydrogens in, say, CH_4; picture the carbon atom in the center of this box, not at any apex, side or edge. Use the 109° template to see how all four hydrogens in methane are equivalent.

Model the Molecules

The progress from methane to other members of the family is easily visualized by considering molecular fragments called *free radicals* (See ECOT Sec. 6.3). A free radical has an unpaired electron and is almost always very reactive. The simplest free radical is a hydrogen atom, formed from the splitting of a hydrogen molecule. Using Lewis dots to express the bonding, we can see how one is formed:

$$H : H \longrightarrow 2\,H\bullet$$

The two electrons shared between the H's are evenly divided; each H ends up with one unpaired electron. Next, let's consider the formation of free radicals from the methane molecule:

$$\begin{array}{ccc} & H & & H \\ & \bullet\bullet & & \bullet\bullet \\ H : & C : & H \longrightarrow & H : C \bullet \;+\; H\bullet \\ & \bullet\bullet & & \bullet\bullet \\ & H & & H \end{array}$$

Again, the electron pair forming the bond between the carbon and the hydrogen is split and each fragment has one of these electrons. To give you some idea how reactive a free radical is, it is helpful to understand how two molecules that are *not* free radicals might react in the gas phase. Take, for example, the reaction of hydrogen, H_2, and chlorine, Cl_2, in the gas phase. First the two molecules must collide. For a collision to be effective, it must occur with the proper geometry. There are an infinite number of ways the two molecules might collide. The collision geometry has to be just right for the reaction to occur. At the same time the *energy* of the collision must be sufficient to bring about reaction.

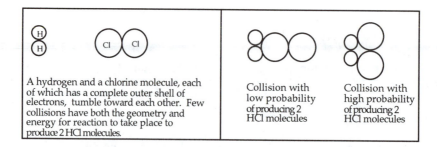

A hydrogen and a chlorine molecule, each of which has a complete outer shell of electrons, tumble toward each other. Few collisions have both the geometry and energy for reaction to take place to produce 2 HCl molecules.

Collision with low probability of producing 2 HCl molecules

Collision with high probability of producing 2 HCl molecules

For these H_2 and Cl_2 molecules there may be a thousand collisions before a fruitful collision occurs. But when two methyl radicals collide they almost always form an ethane molecule on the first collision. Their "hunger" for electrons forces the rearrangement of atoms and electrons into the singly bonded ethane molecule.

$$\underset{\substack{|\\H}}{\overset{\substack{H\\|}}{H-C\cdot}} \;+\; \underset{\substack{|\\H}}{\overset{\substack{H\\|}}{\cdot C-H}} \longrightarrow \underset{\substack{|\\H}}{\overset{\substack{H\\|}}{H-C}}\underset{\substack{|\\H}}{\overset{\substack{H\\|}}{C-H}}$$

The flames in Kekulé's fireplace are evidence of rapid reactions in which a large amount of energy is released. The temperatures are high enough to form free radicals that react with oxygen or other species at a high rate.

ECOT mentioned that alkanes are relatively inert and undergo reactions only with difficulty. If so, why are they so useful in combustion reactions as auto fuel? To answer this, we must consider the conditions inside the auto engine cylinder. The pressure is eight or nine times higher than normal, the temperature is hundreds of degrees above that outside the engine block, and on top of all that, this mixture is further encouraged to react by having a 2500 volt spark discharged through it. This is really drastic treatment. No wonder it explodes.

To appreciate the three-dimensional structure of organic molecules in any depth, you may wish to purchase a model kit. Organic model kits tend to be expensive, however, so you can make models at considerably less expense by buying a bag of gumdrops and a box of toothpicks. Single bonds can be represented by one toothpick, double bonds by two. It is easy to distinguish kinds of atoms by the use of colors; the table below shows a conventional color scheme.

The Gum Drop Organic Model Kit	
ATOM	*COLOR*
H	white (vanilla)
C	black (licorice)
O	red (cherry)
S	yellow (lemon)
N	green (mint)
F, Cl, Br, I	orange (!)

You should make models of all the molecules in Sections 6.5 and 6.6, a selection of models from Section 6.10 and each of the models at the bottom of Table 6.5 of ECOT. Those models from the table should be arranged on a surface as they appear at the bottom of the table and studied to see how they are different. Benefits accruing from this overall exercise include:

- Clearly seeing the three-dimensionality of the tetrahedral bonding of carbon;

- Seeing how a family of alkanes or alkenes, for instance, can be built merely by the insertion of a $-CH_2-$ group in the chain;

- Being prepared when the frequently asked question, "Identify the compound whose model is on display at the front of the class," turns up on a quiz over material treated on this chapter in ECOT.

(An additional advantage is that the Gum Drop Organic Model Kit produces edible organic models. The only drawbacks are that a molecule the size of butane is worth about 100 calories and can lead to dental cavities!)

Organic Nomenclature

The following question is often found on quizzes on organic nomenclature: Name the compound represented by the structure below:

$$CH_3 - \underset{\underset{CH_2CH_3}{|}}{CH} - CH_2CH_2CH_2CH_3$$

The "longest chain" appears to be six carbons in length, with an ethyl side chain on the number 2 carbon, and therefore named 2-ethyl hexane. Not so. In your head, take hold of the ethyl and the butyl ends and straighten the bend.

$$CH_3CH_2 - \underset{}{\overset{\overset{CH_3}{|}}{CH}} - CH_2CH_2CH_2CH_3$$

What appears to be an ethyl side chain is really part of the main chain of a *hep*tane molecule, with a methyl group on the number 3 carbon! The correct name for this compound is 3-methylheptane. Formulas for compounds, especially organic compounds, are sometimes laid out on the page to fit, or to emphasize a reactive site rather than to represent the true geometric arrangement in space. No amount of drawing, sketching or visualizing molecules can substitute for building models of them.

The first few names of the alcohol family are given in ECOT. Note that they have the same prefixes the alkanes have: meth-, eth-, prop-, etc. You should be able to name the first 10 straight-chain alcohols from memory. Then, practice drawing and modeling their structures.

Questions Answered

A Candle Burning... Heat and Light

6.1   Before Wöhler's experiment, all "natural philosophers" thought that organic compounds, including urea (isolated from urine) had to come from living things, not from any inorganic source such as ores, sea water or isolated from the air. Today an organic chemical is one containing carbon.

6.2   4; 8; 10.

6.3   The general formula of an alkane is C_nH_{2n+2}. So, if the alkane is decane, there are 10 carbons and (10 x 2) + 2 = 22 hydrogens.

6.4   Petroleum jelly, mineral oil, lubricating oil, candles, LP gas, paint thinner, etc.

6.5   In butane: carbons: 2, 2, 0; hydrogens: 6, 4, 0.
      in methylpropane: carbons: 3, 0, 1; hydrogens: 9, 0, 1.

6.6   Unbranched, branched, branched.

6.7 $C_{18}H_{38}$

6.8   Isopentane (see structure in Section 6.8)

6.9 The five isomers of hexane are:

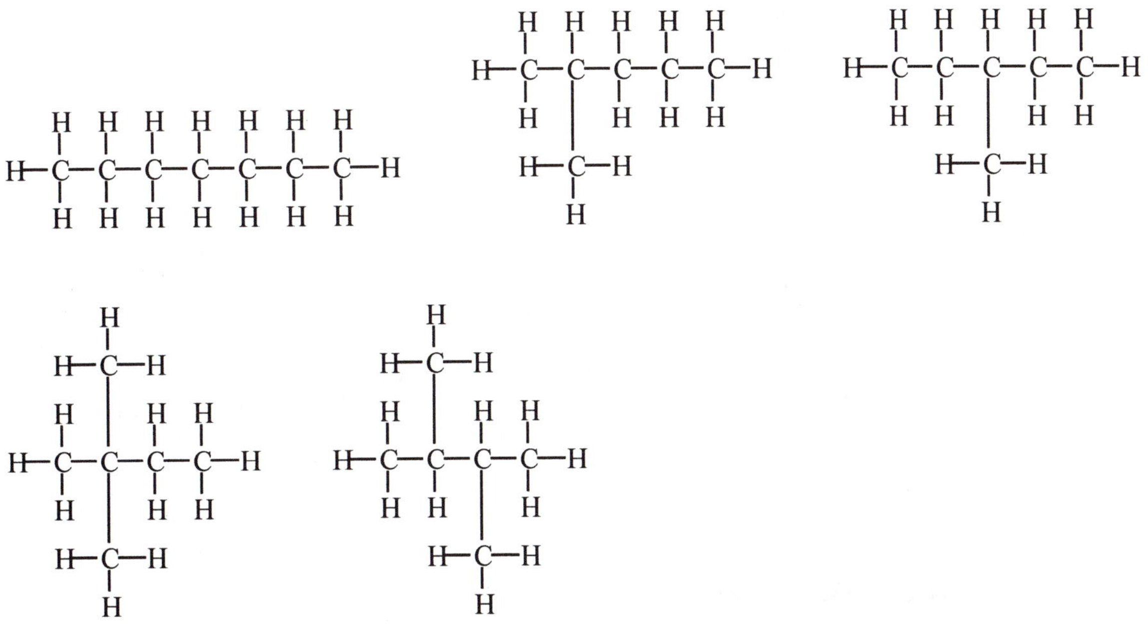

6.10 Not necessarily. An alkene has one double bond, which means that it has 2 fewer hydrogens than an alkane, so its general formula is C_nH_{2n}, but a cycloalkane has the same general formula, because it contains one more C—C bond (and thus that 2 fewer hydrogens) than a straight chain alkane. More would have to be known about the compound before it could be classified as an alkene.

6.11 All aromatic compounds have one or more benzene rings in their molecules.

6.12 $CH_3—CH=CH—CH_3 + H_2 \rightarrow CH_3—CH_2—CH_2—CH_3$

Addition of H_2 to 2-butene also yields butane.

6.14 $C_{25}H_{52} + 38\ O_2 \rightarrow 25\ CO_2 + 26\ H_2O$

Supplementary Exercises

S1. Which of the following compounds have more than one double bond? a. C_2H_4 b. C_3H_8 c. C_5H_8 d. $C_{12}H_{26}$ e. $C_{17}H_{34}$ f. C_9H_{16}

S2. Which is the only proper condensed formula for C_4H_{10}?

a. $CH_3CH_2CH_3CH_2$ b. $CH_2CH_3CH_3CH_2$ c. $CH_3CH_2CH_2CH_3$

S3. Which of the two compounds below are isomers?

a.

$$\underset{\displaystyle CH_3-\overset{\displaystyle O}{\overset{\displaystyle \|}{C}}-CH_2CH_3}{}$$

b.

$$\underset{\displaystyle CH_3-\overset{\displaystyle OH}{\overset{\displaystyle |}{CH}}-CH_2CH_3}{}$$

c.

$$\underset{\displaystyle CH_3CH_2CH_2-\overset{\displaystyle O}{\overset{\displaystyle \|}{CH}}}{}$$

d.

$$\underset{\displaystyle CH_3-\overset{\displaystyle OH}{\overset{\displaystyle |}{CH}}-CH_2CH_2OH}{}$$

S4. Which two compounds below belong to the same chemical family?

a. 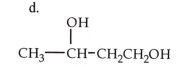 b. c. d.

S5. Which two compounds below are isomers?

a. b. c. d.

S6. Name the following compounds according to IUPAC rules.

a. $CH_2{=}CH_2CH_2CH_2CH_3$

b. $CH_3CH{=}CHCH_2CH_3$

c. $CH_3CH_2CH{=}CHCH_3$

d. $CH_3CH_2CH_2CH{=}CH_2$

S7. With your gumdrop kit, make models of the following molecules. (Or, at least draw them!)

a. 2,2-dimethyl pentane   b. 2-methyl-1-butene   c. ethylbenzene

Where You Might Goof Up...

- Being unable to distinguish between organic and inorganic compounds either by inspection of formulas or in response to an essay question.

- Being unable to write two hundred words on the significance of Wöhler's experiments.

- Not recognizing the name of a compound discussed in Chapter 6 of ECOT whose model is on display.

- Forgetting the names of any of the first ten alkanes.

- Forgetting the root prefixes listed by IUPAC.

- Not picking all isomers of a compound from a list of expanded structural diagrams.

- Inadequately discussing the chemical and electronic nature of free radicals.

- Improperly applying the IUPAC rules when naming compounds.

- Being unable to balance combustion equations for any hydrocarbon up to C_5.

- Erroneously selecting a specified alkane, alkene or alkyne from a list of formulas or expanded structural diagrams.

- Having little familiarity with trivial names allowed by IUPAC such as ethylene, acetylene, isobutene or toluene.

- Insufficiently discussing the bonding associated with aromatic compounds, especially the "smeared" electrons.

- Being unprepared to argue both sides of the greenhouse effect.

CHAPTER 7

Petroleum

THE DRIVING FORCE OF ENERGY

You've changed the oil in your car, and now you have a dilemma. In a pan in your driveway sits a dirty, viscous toxic waste. This workhorse of a lubricant has spent its slick molecules to extend the life of your engine, and, in the process, has become an oxidized, heavy-metal-contaminated heterogeneous mixture of hydrocarbons, various oxidation products, and thick sludge containing water and solids.

What do you do with it? Hide it in your trash where it can go undetected to the sanitary waste landfill with the other 400 million gallons* of waste motor oil improperly disposed of annually? Put it in a gallon bucket and stack it in your garage with all the other gallon buckets containing waste oil, stored until you can think of something else to do with it or until it becomes part of your estate after your death when some heir discovers it?

There's an easy and environmentally friendly way to dispose of the nasty mess. For many years used oil has been collected from service stations, cleaned, and marketed as recycled oil or heating oil. Now it is becoming a starting material for gasoline, replacing some of the crude oil required for the refining process. Refining cracks the large hydrocarbon molecules of used motor oil, and the gasoline product is the same as that from crude oil refining. The potential for decreasing our dependence on new crude makes this an exciting development, but it depends on availability of large quantities of used oil.

Recycling centers are becoming easier to find as communities set up collection sites where used oil can be dropped off. Find a drop-off site in your area.

Chapter Overview

- An experiment designed to demonstrate the most-used procedure in the petroleum industry is recommended.

- Our dependence on oil is emphasized.

- The automobile engine and its fuel are described.

- The physical and chemical properties of the organic compounds described in the previous chapter are discussed.

- Knocking and the solution to this problem are covered.

- Octane ratings are clarified.

- The avoidance of lead in modern gasolines and substitutes for this metal are mentioned, and catalytic converters are examined.

- Cracking and reforming, the rearrangement of atoms in organic molecules to get more fuel out of each barrel of oil, are described.

- Alternative automobile fuels are listed, with commentary.

---

* This is *35 times* the volume of the Exxon Valdez oil spill in Prince William Sound. Ron Cogan, "Car Care: Recycling Lubricants," *Motor Trend*, September, 1990, 122.

For Emphasis

Molecular Structure and Boiling Point

After examining Table 7.1 of ECOT you might wonder, "Why does the boiling point increase with chain length?" As chain length increases, the molecule gets heavier. As the molecule gets heavier, it takes more energy (higher temperature) to lift it out of the body of the liquid; forces of attraction (relatively weak, compared to those associated with chemical bonds) must be overcome as well. The chains move in close proximity to each other in the liquid, weakly attracting each other. The longer the chain, the greater these interactions. However, when comparing straight chain molecules of different length, the dominant influence will be the mass of the molecule rather than attractive interaction.

Now consider the effect of chain branching on boiling point. Table 7.2 of ECOT shows how boiling point changes with chain branching. Why is it that side chains lower the boiling point for molecules having the same molecular weight? From what is said in the paragraph above, we might expect that the boiling point should be the same or nearly the same if the molecular weights are the same. Table 7.2 shows that's clearly not the case.

To see how branching makes a difference, assemble two gumdrop models of butane and put them side by side with as many H's touching as you can. The electron cloud of one hydrogen is attracted by the nucleus of the hydrogen of the molecule lying beside it and *vice versa*. Now change the gumdrops so that you have two isobutanes instead. Put the two molecules side by side with as many H's touching as you can. No matter how you orient the two models, you cannot get as many hydrogens to touch each other as you did with the straight chain butanes. There just is not as much interaction between branched chain molecules as straight chain ones. There are fewer hydrogens able to get close to each other.

A worse case is shown by comparing pentane, isopentane and neopentane (Table 7.2 in ECOT). As we move from the straight chain pentane through the single methyl side chain isopentane to the doubly branched neopentane, it is like moving from a pencil to a tennis ball. Two pencils can be placed in such a way that they touch along their whole lengths; two tennis balls can make contact at only a very small point. If we imagine two pencils that can attract each other only while touching and two tennis balls that can attract only while touching, we can see the difference easily. This analogy holds well with molecules. The straight ones attract each other and need more energy to escape (higher boiling point), while the branched chain molecules do not attract each other as much and can escape at a lower temperature (need less energy) than the straight chain ones.

Compression Ratio and *Knocking*

An understanding of compression ratios is important in automotive and combustion engineering. Calculation of a compression ratio is easy when the data are at hand. Turn to Figure 7.2 of ECOT showing a cylinder at the beginning and at the end of the compression stroke. Suppose this is one cylinder of a four-cylinder 2.0 liter automobile engine. At the beginning of the stroke, the volume of the air/fuel mixture would be 500 cc; this is volume "A" in the diagram. Suppose further that the volume of the air/fuel mixture at the top of the stroke just before detonation is about 55 cc, which is volume "B" in the diagram. The compression ratio of this engine would then be

$$\frac{\text{Volume A}}{\text{Volume B}} = \frac{500 \text{ cm}^3}{55 \text{ cm}^3} = 9.1$$

Automotive engineers seek high efficiency by going to higher and higher compression ratios, because a higher compression ratio means higher ignition temperatures which means higher efficiency. As described in ECOT, a ceiling is placed on gasoline engine design because of knocking. As the compression ratio increases, so does the engine's penchant for pre-ignition, or knocking. In an attempt to get around this, the diesel engine was invented, using a fuel of much higher boiling range (see Table 7.4 of ECOT) and a much higher compression ratio. The higher compression ratios lead to very high combustion temperatures and increased efficiency; the diesel fuel is so well-behaved that pre-ignition doesn't occur even at diesel compression ratios of 16 or 18 to 1.

Catalytic Cracking

There are three rules of thumb associated with predicting the products of a catalytic cracking process:

- A large molecule usually splits near its middle when cracked.

- Starting with a large alkane, one product molecule will be an alkane and the other an alkene.

- The double bond of the alkene will be in the #1 position.

We emphasize that these are rules of thumb, and mixtures of compounds will always be found as products of cracking. These rules become less effective as the temperature of the cracking process is raised. Catalysts are used to make reactions go at lower temperatures and to choose one reaction path over another.

Research in catalysis is of the highest priority in the petroleum industry. In some phases of cracking and reforming, the engineer can almost "dial a product" with the selection of the correct catalyst. In the example of selectivity below, gaseous ethyl alcohol is heated in the presence of two different catalysts, aluminum oxide and copper metal.

$$C_2H_5OH \xrightarrow{Al_2O_3} C_2H_4 + H_2O$$
$$C_2H_5OH \xrightarrow{Cu} CH_3CHO + H_2$$

In the first case, with Al_2O_3 as catalyst, the products are ethylene and water, whereas when Cu is the catalyst, acetaldehyde and hydrogen are formed. The equations as written above are according to convention in that the catalyst is written above the arrow. Specifying products in reforming processes by selection of catalyst can be done as well.

Using catalysts is an effective way of accelerating reactions as well as selecting one set of products over another. Since most chemical reactions release heat during their progress, catalyzed reactions produce heat so rapidly that temperatures rise quickly, sometimes so quickly that brush or grass fires have been started under catalytic converters of cars left idling on the side of the road.

Questions Answered

Petroleum and Strong Tea...In ECOT, the author writes: "Fossil fuels are fuels formed from the partially decayed animal and vegetable matter of living things that inhabited the earth in eras long past." Thus, wood from trees being cut down now (no matter by whom) are not fossil fuel.

7.1   The boiling point.

7.2   a. Compression and power; b. intake and exhaust; c. none.

7.3   Octane, with the highest molecular weight and straight chain has the highest boiling point; 2,3-dimethylbutane has the most branching of the six-carbon molecules and will have the lowest boiling point.

7.4 Knocking, the sound of explosions in the engine, is a rapid pinging that can come from an engine under sudden acceleration or heavy load. It can result from preignition, in which the gas/air mixture ignites before the piston has reached the point where the spark plug usually fires. It can also be caused by ignition beginning in several different spots in the cylinder simultaneously.

7.5 The "octane" with an octane rating of 100 is trimethylpentane. Note that it has 8 carbons, hence the name octane, but it is highly branched. Octane with an octane rating of –20 is a straight-chain molecule. The two structures are isomers.

7.6 $C_8H_{20}Pb$.

7.7 The quantity of unburned hydrocarbons and CO are reduced by catalytic converters. Nitrogen and sulfur oxides are unaffected.

7.8 Lead is a poisonous pollutant to living beings as well as to catalytic converters.

7.9 MTBE, because it has a higher octane rating.

7.10 Chain-branching.

7.11 According to the rules of thumb stated above, cracking usually produces an alkane and an alkene. Since a six-carbon alkene is already one product, a three-carbon alkane should be the other product, hence propane: $CH_3CH_2CH_3$.

7.12 Isopentane and neopentane; benzene and hydrogen.

Supplementary Exercises

S1. The volume of a piston in a racing car at the bottom of the power stroke is 1250 cc and 120 cc at the top of the compression stroke. Calculate the compression ratio of this piston.

S2. Tetraethyl lead (TEL) is 64% Pb and has a density of 1.653 g/mL. Assume that tetraethyl lead was still used as an additive in the year 2000 at the rate of 1 teaspoonful (about 3.5 mL) of TEL per gallon of gasoline. Using the number of gallons of gasoline burned worldwide in 2000, and assuming that all the Pb in the environment would have come from lead in gasoline,

a. calculate the total mass in grams of Pb that would have entered the environment during that year.

b. How much lead is this per day?

c. If there were 248 million people in the U.S. in 2000, how many grams of Pb per person would that be for each day?

S3. Write the condensed structure of the isomer of C_7H_{16} which has the largest possible number of methyl groups.

S4. When the straight-chain alkane, $C_{11}H_{24}$, is cracked, one of the products is $CH_3CH_2CH_2CH_2CH_2CH_3$. Suggest the likely other product by name and draw the structural diagram for that molecule.

S5. When the straight-chain $C_{10}H_{22}$ alkane is cracked, one of the products is $CH_3CH_2CH_2CH=CH_2$. Suggest the likely other product by name and draw its structural diagram.

S6. Write the condensed structure for the hydrocarbon which, when cracked, produces $CH_2=CHCH_2CH_3$ and $CH_3CH_2CH_2CH_2CH_3$.

Where You Might Goof Up

- Not being able to describe the distillation process.

- Incorrectly predicting which of two hydrocarbons would more rapidly distill from a mixture even though you were given the boiling points.

- Being unable to distinguish between a two-stroke engine and a four-stroke engine.

- Inaccurately listing the four strokes of the typical automobile engine.

- Erroneously describing the purpose of each of the four strokes of the typical automobile engine.

- Writing 150 words with accompanying diagrams on the subject of "knocking" which fail to convince your instructor that you really know what you are talking about.

- Not distinguishing between octane the compound and octane the fuel rating.

- Being unable to describe the two different ways of determining a fuel's octane rating.

- Improperly predicting the products of cracking given typical starting materials.

- Improperly predicting the products of reforming, isomerization, cyclization and aromatization, given typical starting materials.

- Listing an inadequate number of attributes of a catalyst.

- Omitting any mention of electricity, ethanol, methanol and natural gas when asked to discuss alternate fuels for automobiles.

CHAPTER 8

Working with Chemistry

FOOD, FUEL, AND ENERGY

The old air conditioner wheezed and clunked with a metal-screeching regularity. Outside the sun was beginning its relentless tour of the morning sky, and heat waves rose from concrete that had barely cooled overnight. "115 in the shade today, folks!" shouted a disembodied voice from the clock radio. "Stay inside and keep cool. It's another scorcher—over 110 for 30 days in a row." Already sweaty and uncomfortable, Julia arose and looked outside. Tree skeletons stood against the hazy, hot sky, silent witnesses to the ravages of the five-year drought. What had once been a soft, verdant front yard was a dust bowl of brown dirt and pebbles. A hot breeze kicked up a dust devil that whirled for a moment and then settled, its energy expended. Julia grimaced and turned away from the window to get ready for her first day at a new job. "Welcome to Denver," she muttered.

Could this happen? In 2002, a year-long drought in the Front Range of Colorado, the mountain range that Denver abuts, has forced water rationing in communities that already face a diminishing water supply and increased population. Denver's "brown cloud" is legendary, partially caused by a weather phenomenon produced by the close proximity of the Rocky Mountains and increasingly enhanced by unburned hydrocarbons from automobile travel and industry along the Front Range. A study published by the Regional Air Quality Control in 2000 showed that gas and diesel exhaust contributed 38% of the compounds in the brown cloud. We know from ECOT that gas and diesel exhaust contain unburned hydrocarbons, sulfur and nitrogen oxides, carbon dioxide, and ozone. The combination of drought with fossil fuel combustion products that contribute to the greenhouse effect has the potential to change our climate. What can we do about it?

Chapter Overview

- Chemical formulas are translated into English and back again into formulas.

- A demonstration requiring no extra lab equipment is encouraged.

- Energy, both potential and kinetic, are defined.

- The calorie is defined, and a calorimetry experiment is discussed.

- The relationship between work and energy and the *joule* as the unit of work and energy are introduced.

- Energy-releasing and energy-consuming reactions are used to illustrate *exothermic* and *endothermic* processes, respectively.

- Metabolism is defined, and the human body is compared to a car engine.

- The energy equation, Energy In = Energy Out + Energy Stored, is given.

- The three ways the human body uses energy are described, and the three classes of chemicals that provide the bulk of our food supply are discussed.

- The sun as the source of almost all of our energy and its relationship to fossil fuels is outlined.

- CO_2, the greenhouse effect, and global warning are described.

For Emphasis

Heat, Work and Thermal Energy

Heat is energy on its way to someplace. Heat is thermal energy flowing into or out of a body; thermal energy is energy residing in a body because its temperature is above absolute zero. As this chapter is studied, the following points should be kept in mind:

- heat can be exchanged for work, and vice versa.

- heat always flows from a warm body to a cool one.

- the potential for heat to be produced resides in chemical bonds and can be released during chemical reactions.

How to Approach Calorimetry Problems

The easiest way to approach a calorimetry problem is intuitively. Suppose you heat 100 g of water from 20° C to 80° C. How much heat (in calories) was added to the water?

Start with the definition of a calorie: 1 calorie is the quantity of heat necessary to raise the temperature of 1 g of water 1° C. So, if we had 1 g of water and raised the temperature 60°, we would have used 60 cal of heat. Since we raised 100 grams through that temperature interval, we used 100 x 60 or 6000 cal of heat.

Reviewing Units Cancellation

Because the author of ECOT uses unit cancellation as a problem solving tool in this chapter, we should take a few paragraphs to do the same thing in some detail. Bear in mind that this is not the only way to solve problems. But it can be a convenient tool that can be applied to many different types of problems. Take a problem like the Question at the end of Section 8.5 in ECOT. Read the question, and then, if you are not feeling comfortable with units cancellation, continue reading below.

When you are using units cancellation, keep in mind what unit the answer must have. We know that a joule = watt x second, or $1 = \left(\dfrac{watt \cdot second}{joule} \right)$. Isn't it also true that $1 = \left(\dfrac{joule}{watt \cdot second} \right)$? Remember, in Chapters 1 and 2 of this *Guide* we examined relationships like this one which make true statements either way they are written. In the same way we can say that the ratio of joules to calories is $\dfrac{4.2 \ joules}{1 \ cal}$, or the ratio of calories to joules is $\dfrac{1 \ cal}{4.2 \ joules}$. We can use these ratios (which, of course we have been doing already throughout this *Guide*) to give us an answer in the units we seek.

Now, back to the Question at the end of Section 8.5. We want an answer in "hours," so we write down what we are starting with and what we want to finish with. We leave lots of space in between to fill with ratios.

Start with:

4.0 x 10⁶ joules   =   hours.

End with:

$$\text{hours} = \text{hours}$$

We start multiplying by ratios that will successively cancel the units we don't want and lead us to the units we want in the answer (hours), which have to end up in the *numerator* on the left. We can invert the ratios as needed; our 40-watt bulb can be used as 40/1 or 1/40 as the units demand. Try it.

If the answer you get is not "28 hours," turn to the problem solution in "Questions Answered," and look at the setup. Then work the Supplementary Exercises.

Questions Answered

8.1   Paraffin (a hydrocarbon); the candle stops burning; the animal dies

8.2   Let's start with potential energy. The greatest amount of potential energy is at the point where there is no motion. The point at which the child has no motion is at the ends of the arc the swing makes. In other words, it's at the point that the child's motion actually stops and changes from forward to backward or backward to forward. So those two extremes are the point of greatest potential energy. What is the point of greatest kinetic energy? As the swing moves from one of the ends of the arc, it accelerates as it approaches the midpoint of the arc (when the ropes holding the swing are perpendicular to the ground). As it passes through the midpoint, it slows until it stops at the top of the arc. The greatest kinetic energy, then, is at the midpoint of the swing's arc.

8.3   Copper atoms are moving more briskly in the hot copper than in the cold.

8.4   Let's approach this problem intuitively, as in the Example in this section (8.4). We know that 1 calorie of heat (work) raises the temperature of 1 gram of water 1 °C. The Count had 12 kg or 12,000 g of water. If his boring had raised the temperature of the water 1 °C, he would have done 12,000 calories of work. As it was, he raised the temperature of the water 80 °C (100° - 20°), so the amount of heat produced (or work done) was equal to 12,000 x 80 or 960,000 calories (9.6×10^5 cal). The problem asks us for an answer in kilocalories. Since 1 kilocalorie is 1000 cal, the Count did 960 kcal of work.

You can also approach this mathematically:

$$12 \text{ kg} \left(\frac{1000 \text{ g}}{1 \text{ kg}} \right) \left(\frac{1 \text{ cal}}{1 \text{ g} \cdot 1 \text{ °C}} \right) (80 \text{ °C}) \left(\frac{1 \text{ kcal}}{1000 \text{ cal}} \right) = 960 \text{ kcal}$$

8.5   The first question is the same problem as 8.3 above, except that to answer it we have to convert the calories we obtained to joules. To convert calories to joules, multiply calories (*not* kilocalories) by 4.2 joules/cal. So the number of joules of work would be

$$9.6 \times 10^5 \text{ cal} \left(\frac{4.2 \text{ joules}}{\text{cal}} \right) = 4.0 \times 10^6 \text{ joules}$$

We can easily work the second question intuitively, as we did 8.3, but it is quickly worked by units cancellation, as well.

$$4.0 \times 10^6 \text{ joules} \left(\frac{\text{watt} \cdot \text{second}}{\text{joule}} \right) \left(\frac{1}{40 \text{ watts}} \right) \left(\frac{1 \text{ min}}{60 \text{ sec}} \right) \left(\frac{1 \text{ hr}}{60 \text{ min}} \right) = 28 \text{ hr}$$

For more on units cancellation, see Appendix C in ECOT.

8.6 If you draw the structures of the first alkanes (or simply turn to Table 6.1, p. 127 in ECOT), you will see that the number of bonds in the series increases by 3 with each successive alkane. So, methane has 4 bonds, ethane 7, propane 10, butane 13, and hexane 16. Since pentane releases approximately 845 kcal per 6.02×10^{23} molecules, then hexane will release 53(3) kcal more, or a total of 1,004 kcal of energy.

8.7 Eating a meal.

8.8 a. 100 cal; b. 1×10^{-4} Cal.

8.9 Use units cancellation to solve the parts of this problem.

a. Start by calculating the number of calories per person per year.

$$\left(\frac{200 \text{ Cal}}{\text{hr}}\right)\left(\frac{24 \text{ hr}}{\text{day}}\right)\left(\frac{365 \text{ days}}{\text{yr}}\right) = 1,752,000 \text{ or about } 1.8 \times 10^6 \text{ Cal} \quad \text{per person}$$

Now calculate the energy released through exercise by all the people on the earth in a year:

$$\left(\frac{1.8 \times 10^6 \text{ Cal}}{\text{person}}\right) 6 \times 10^9 \text{ persons} = 1.1 \times 10^{16} \text{ Cal}$$

b. One person is expending one Calorie (1 kcal) per hour per kg of body weight in basal metabolism. For an average body weight of 50 kg, for the 24-hour period, the total expenditure would be

$$24 \text{ hours} \left(\frac{1 \text{ Cal}}{\text{hour} \cdot \text{kg}}\right) 50 \text{ kg} = 1200 \text{ Cal}$$

Now determine the energy output of one person for a year:

$$\left(\frac{1200 \text{ Cal}}{\text{day}}\right)\left(\frac{365 \text{ days}}{\text{yr}}\right) = 4.4 \times 10^5 \text{ Cal/yr}$$

For all the humans on earth, the total would be

$$\left(6 \times 10^9 \text{ persons}\right)\left(\frac{4.4 \times 10^5 \text{ Cal}}{\text{person}}\right) = 2.6 \times 10^{15} \text{ Cal}$$

c. The total energy output, ignoring SDA, would be $1.1 \times 10^{16} + 0.26 \times 10^{16}$ or 1.4×10^{16} Cal/year.

8.10 a. We add up the contribution of each food category:

13 g of protein x 4 Cal/g =	52 Cal
15 g of carbohydrate x 4 Cal/g =	60 Cal
5 g of fat x 9 Cal/g =	45 Cal
total =	157 Cal

b. The soup contains 45 Cal/157 Cal x 100 = 29% fat.

c. This 50 kg individual burns 1 Cal/hour for every kg of body weight, so that is 50 Cal/hour in basal metabolism activity. The chicken noodle soup would support just over three hours of this activity.

8.11 a. Wind energy; b. nuclear or geothermal energy.

8.12 The nuclear energy produced on the sun drives photosynthesis. This energy is transformed into chemical energy in plants.

8.13 Sources: fossil fuels, animal metabolism; Sinks: atmosphere, plants, surface waters.

8.14 Venus is closer to the sun, and it contains 95% CO_2 in a very dense atmosphere, in which the greenhouse effect is the major reason for its high temperature. Mars, on the other hand, which has a much *less* dense atmosphere than Earth, also contains 95% CO_2, but it is farther from the sun than Earth and thus loses heat quickly.

8.15 The temperature rose about 2°C in the last quarter of the 20th century. This means that in the year 2100, the average temperature would be 2°C higher than the average temp in 2000.

Supplementary Exercises

S1. How many kcal of energy are released in burning 1.0 mole of nonane?

S2. What will be the final temperature of water when 30 g of water at 60 °C are mixed with 30 g of water at 30 °C?

S3. What percentage of energy comes from fat in a salad which is 37% fat, 45% protein and the rest carbohydrate?

S4. How long would it take a 70-kg person to walk off that extra piece of pizza and a second malted milkshake?

S5.   How long would it take a burning 250-watt bulb to heat 3.5 kg of water from 25 °C to 53 °C?

Where You Might Goof Up...

- Being unable to define energy, potential energy, and kinetic energy, and failing to be able to describe the many forms of energy.

- Not being able to write a paragraph about Count Rumford's boring experiment.

- Forgetting that heat lost = heat gained and that heat and work are interchangeable.

- Failing to distinguish between calories, kilocalories, and Calories.

- Miscalculating energy exchanges with calorimetry data.

- Confusing exothermic and endothermic reactions and not being able to compute the quantity of energy released or absorbed in a reaction.

- Being unable to list the three main macronutrients: fat, protein and carbohydrate.

- Forgetting that Energy In = Energy Out + Energy Stored.

- Incompletely discussing the specific actions and the rate of energy consumption for exercise, SDA and basal metabolism.

- Being unable to remember the quantity of energy per gram provided by each macronutrient and incorrectly computing their percentages in a sample of food.

- Confusing photosynthesis and respiration.

- Inadequately relating solar energy to fossil fuels.

- Writing an essay on global warming and leaving out a description of the four most important greenhouse gases produced by human activity.

CHAPTER 9

Arithmetic of Chemistry

CONCENTRATING ON POLLUTION

The discussion at the Ridgeton town meeting was heated. "I don't want chemicals in my water," a man shouted. "Why should my kids have to drink water that has anything in it?"

"I drink bottled water from Artesian Springs over in Randolph County. Now, that's good, pure water!" said a young woman. Some people nodded their heads. "Yeah," agreed her neighbor. "If we had water like that I'd be happy."

Ridgeton's water chemist, Dr. Hunter, saw her chance. She stood up.

"Ladies and gentlemen, if you will focus your attention on the screen, I'd like to address this issue of water purity." On the screen she projected a row of figures, keeping the lower two-thirds of the display covered.

Water Analysis, ppm

	Arsenic	Barium	Calcium	Chloride	Lead	Magnesium	Nitrate	Sodium	Sulfate	Zinc	Dissolved Solids
Ridgeton	0	0.02	75	175	0.01	20	7	20	170	4	225

"You will recognize Ridgeton's water analysis published in the Weekly News." There were nods and murmurs from the audience. "Now, focus your attention on this next row." She uncovered another row of numbers. "This is the analysis of the Artesian Springs water some of you have been buying."

	Arsenic	Barium	Calcium	Chloride	Lead	Magnesium	Nitrate	Sodium	Sulfate	Zinc	Dissolved Solids
Art. Springs	0.02	0.05	100	200	0.01	25	3	45	225	1	450

There was shocked silence for a moment.

"You mean bottled water is worse than town water?" someone asked. "There's arsenic in it!"

"It's hard to ascribe 'better' or 'worse' to these figures, when we are talking about such minute amounts," Dr. Hunter responded. "Instead, compare these figures to maximum contaminant levels set by the Safe Drinking Water Act and the EPA." She uncovered another row of numbers. "As you can see, the water from both sources meets the standards in every category. Some of the common minerals which are part of our diet, such as calcium and magnesium, have no set standards in drinking water."

	Arsenic	Barium	Calcium	Chloride	Lead	Magnesium	Nitrate	Sodium	Sulfate	Zinc	Dissolved Solids
Standard	0.05	1	—	250	0.05	—	10	160	250	5	500

It was quiet in the auditorium as Dr. Hunter spoke about the importance of vigilance where nitrate from agricultural runoff was concerned. And the town council offered free water analysis to anyone worried about lead contamination from plumbing in older homes.

Chapter Overview

- A dilution experiment is outlined which helps to frame the question, "How much pollution is tolerable?" Trying this at home would be instructive.

- A quantitative discussion of how atoms and molecules react with each other is conducted.

- The standard way chemists count their particles, the *mole*, is introduced and manipulated.

- Using balanced chemical equations to predict how much of this will react with how much of that is demonstrated.

- Concentration units of molarity, percentage composition, and parts per million are defined and some calculations done.

- The purity of our drinking, cooking, swimming, and washing water is characterized.

- Pollution is portrayed as "matter out of place."

For Emphasis

As we have seen in Chapter 9 in ECOT and in the scenario above, the questions that are most meaningful in chemistry are not always the *what* or *whether* questions, but are often the *how much* questions. And when it comes to our drinking water, we want to know not only how much of any substance besides H_2O is there, but how much is harmful. And our *how much* questions extend well beyond questions of pollution.

What an Equation Says

To be quantitative about chemistry, we need balanced equations. Further, we must know how to read them. They are, in a very real way, chemical sentences in English, complete with subject, verb and predicate.

Let's consider an example:

$$C + O_2 \rightarrow CO_2$$

The quick way to read this is to say, "Carbon plus oxygen yields carbon dioxide." The arrow is sometimes read as *gives, becomes, equals, produce, react to become.* There is important quantitative information as well in every chemical equation.

- One atom of carbon and one molecule of oxygen give one molecule of carbon dioxide.

- One mole of carbon plus one mole of oxygen will produce one mole of carbon dioxide.

- Twelve grams of carbon and thirty-two grams of oxygen equal forty-four grams of carbon dioxide.

Let's take another example:

$$2\,C + O_2 \rightarrow 2\,CO$$

Now we can say that

- Two atoms of carbon and one molecule of oxygen give two molecules of carbon monoxide.

- Two moles of carbon plus one mole of oxygen will produce two moles of carbon monoxide.

- 24 grams of carbon and 32 grams of oxygen equal 56 grams of carbon monoxide.

But what if we only have 12 grams of carbon? How many grams of carbon monoxide can form? Common sense (and a little chemistry) tells us that if we have half the amount of carbon we would expect to get half the amount of product, or 28 g of CO. This type of problem is called *stoichiometry*, the chemistry of how much of this reacts with how much of that to produce how much product; or, the study of the quantities of reactants and products in chemical reactions.

How Much of This reacts with That ?

We are now in a position to calculate quantities of reactants and products in a very general way. If you master the technique outlined below, there is no stoichiometry problem that you can't solve. The author of ECOT has used a five-step approach and we will do the same. After a while you will probably note that you are doing part of each problem in your head and the number of steps you go through on paper may decrease.

Let's go back to the question asked at the end of the last section. How many grams of carbon monoxide will be formed from 12 g or carbon? (We know that oxygen is one of the reactants, but we always assume, since it is not mentioned, that there is enough oxygen for the reaction to proceed to use up all 12 g of carbon.)

Step 1: Write a balanced chemical equation.

$$2\,C + O_2 \rightarrow 2\,CO$$

Step 2: Start with the *known weight* of the chemicals taking part in the reaction. Write that weight above the chemical in the equation. Put an "x" over the chemical we are looking for. This helps you keep track of the quantities.

$$\overset{12\,g}{2\,C} + O_2 \rightarrow \overset{x}{2\,CO}$$

Step 3: How many *moles* of the given substance is this?

$$\text{mol of C} = 12\,g\,C \left(\frac{1\,\text{mol C}}{12\,g\,\text{of C}} \right) = 1\,\text{mol of C}$$

Step 4: Look at the mole relationship given by the coefficients of each substance and, using the moles of reactant we calculated in Step 3, calculate the *number of moles of product* we can expect.

According to the equation, 2 moles of C produce 2 moles of CO. We have 1 mole of C, so we can expect to produce 1 mole of CO.

Step 5: Using the *moles* of substance we produce (answer to Step 4), calculate the *weight* of that substance.

$$1\,\text{mol of CO} = 12\,g + 16\,g = 28\,g\,CO$$

Check this answer with our "common sense" answer in the section above. It's the same.

Now, if we look carefully at the steps above, we can see that the first step is always writing the balanced equation. We can't work the problem without doing that. The next four steps can be illustrated graphically. Imagine walking through a stoichiometry problem. Always begin with what's given (where else *could* we begin?). So we begin with 12 g of carbon and walk through the calculation in Steps 2-5 to reach the answer to the question.

$$\text{Start} \overset{12\,g}{\underset{1\text{ mole}}{2C}} + O_2 \rightarrow \overset{28\,g \text{ Finish}}{\underset{1\text{ mole}}{2\,CO}}$$

If we know the mass of one substance in the equation, we can find the mass of any other element or compound. We could be given the mass of CO and asked to find the amount of C that would have to be used to produce that mass of CO. In that case we would start our walk through the equation on the right with CO and end up on the left with C.

So now let's work some problems.

Example 1. What weight of sulfur dioxide, SO_2, results when 36 g of sulfur are burned in air?

Solution:

Step 1: Write the balanced chemical equation: $S + O_2 \rightarrow SO_2$.

Step 2: We are given 36 g of sulfur. Show that on your equation. $\overset{36\,g}{S} + O_2 \rightarrow \overset{x}{SO_2}$.

Step 3: The weight of one mole of sulfur is 32 g (atomic weight in grams). Calculate the moles of S given:

$$\text{mol of S} = 36 \text{ g of S}\left(\frac{1 \text{ mol of S}}{32 \text{ g}}\right) = 1.12 \text{ mol S}$$

Step 4: From the equation we can see that 1 mole of S produces 1 mole of SO_2. Therefore, 1.12 moles of S will produce 1.12 moles of SO_2.

Step 5: Calculate the number of grams of SO_2 in 1.12 moles. To do so we need the molecular weight of SO_2, which is 32+ 2 (16) = 64 g/mole.

$$\text{grams of } SO_2 = 1.12 \text{ mol } SO_2\left(\frac{64 \text{ g } SO_2}{\text{mol } SO_2}\right) = 72 \text{ g}$$

Go back and reread the problem to see if your answer has answered the question asked.

Example 2. In the presence of the catalyst vanadium pentoxide, sulfur will oxidize to sulfur trioxide, SO_3, given an adequate amount of oxygen. The *unbalanced* equation for the process is:

$$S + O_2 \xrightarrow{V_2O_5} SO_3$$

How many g of O_2 are needed to produce 150 g of SO_3? (A catalyst is used to speed up a reaction. It is unchanged in the reaction and is often written over the arrow to indicate its presence.)

Solution:
Step 1: Write the balanced chemical equation. Because there are 2 oxygens in an oxygen molecule but 3 oxygens in an SO_3 molecule, choose 6 as the lowest number

---

of oxygen atoms there can be on one side of the equation. Remembering that the only numbers that we can use must be coefficients, we write:

$$S + 3O_2 \xrightarrow{\text{V}_2\text{O}_5} 2SO_3 \text{ (unbalanced)}$$

Count the S atoms and see that there is 1 S on the left, and there are 2 S atoms on the right. Place a 2 before the S on the left, and the equation is balanced.

$$2S + 3O_2 \xrightarrow{\text{V}_2\text{O}_5} 2SO_3 \text{ (balanced)}$$

Step 2: We are given 150 g of SO_3, and we are to find the weight of O_2 that would be used to produce that weight of SO_3. Show all that on your equation.

$$\overset{x}{2S} + 3O_2 \xrightarrow{\text{V}_2\text{O}_5} \overset{150\text{ g}}{2SO_3}$$

Step 3: The weight of one mole of SO_3 is 32 + 3 (16) = 80 g/mole. Determine how many moles of SO_3 we have.

$$\text{mol of SO}_3 = 150 \text{ g} \left(\frac{1 \text{ mol SO}_3}{80 \text{ g}} \right) = 1.875 \text{ mol}$$

Step 4: The mole relationship of SO_3 to O_2 is 2 to 3. Stated another way we can say that 2 moles of SO_3 are produced by 3 moles of O_2. In other words, to produce 1 mole of SO_3, we would have to have 3/2 or 1.5 moles of O_2. (Be sure to convince yourself of that before reading further.)

$$\text{mol of O}_2 \text{ needed} = 1.875 \text{ mol SO}_3 \left(\frac{3 \text{ mol O}_2}{2 \text{ mol SO}_3} \right) = 2.81 \text{ mol of O}_2$$

Notice that the mole ratio fraction is set up so that moles of SO_3 will cancel. If you accidentally invert the fraction, the units won't cancel, and you know you have made an error. This gives you an instant check on whether the problem is set up correctly.

Step 5: The final step is to find grams of O_2.

$$\text{grams of O}_2 = 2.81 \text{ mol O}_2 \left(\frac{32 \text{ g}}{1 \text{ mol O}_2} \right) = 90 \text{ g of O}_2$$

Let's do one more example, in which we need to calculate how much of one reactant is needed for a given amount of another.

Example 3. The balanced equation below describes the burning of butane in oxygen.

$$2C_4H_{10} + 13O_2 \rightarrow 8CO_2 + 10H_2O$$

How many grams of oxygen are needed to burn completely 250 grams of butane?

Solution:

Step 1: We are given the balanced chemical equation.

Step 2: We have 250 grams of butane, and we want to find the weight of oxygen necessary to burn this amount of butane completely.

$$\overset{250\ g}{2\,C_4H_{10}} + \overset{x}{13\,O_2} \rightarrow 8\,CO_2 + 10\,H_2O$$

Step 3: How many moles of butane do we have?

The molecular weight of butane = 4 (12.0) + 10 (1.0) = 58 g/mole.

$$mol\ of\ C_4H_{10} = 250\ g\left(\frac{1\ mol}{58\ g}\right) = 4.31\ mol$$

Step 4: The mole ratio of butane to oxygen is 2 to 13.

$$mol\ of\ oxygen = 4.31\ mol\ C_4H_{10}\left(\frac{13\ mol\ O_2}{2\ mol\ C_4H_{10}}\right) = 28.0\ mol$$

Step 5: $grams\ of\ O_2 = 28.0\ mol\left(\dfrac{32\ g}{1\ mol}\right) = 896\ g.$

Avogadro's Number and the Mole

The idea of a *mole* is not hard to understand if you restrict yourself to thinking of it is a *number*, a very large number, but still just a number. Chemists need this quantity called a mole because equations speak to us in numbers of atoms or molecules rather than grams. Atoms and molecules do not react together in gram-to-gram or pound-to-pound quantities. They do so in one-to-one, one-to-two or other simple integer relations of atom-to-atom or molecule-to-molecule amounts.

The key to the size of a mole for an element or compound is the atomic or molecular weight; 12 g of carbon is one mole of carbon, 39.1 g of potassium, K, is one mole of potassium, 58.5 g of NaCl is one mole of sodium chloride, 342 g of sucrose, $C_{12}H_{22}O_{11}$, is one mole of sucrose. The magic of thinking in terms of a mole is that all the quantities given in the previous sentence contain *the same number of particles*. That number, as you read in ECOT, is 6.02×10^{23}. The units of this quantity are usually molecules/mole or atoms/mole. The number itself is referred to as Avogadro's Number and is treated arithmetically just like any other number.

Avogadro's Number in Action

- 12g of C (one mole) contains 6.02×10^{23} atoms of C.
- 6g of C (0.5 mole) contains 3.01×10^{23} atoms of C.
- 39.1g of K (one mole) contains 6.02×10^{23} atoms of K.
- 44g of CO_2 (one mole) contains 6.02×10^{23} molecules of CO_2.
- 30g of NO (one mole) contains 6.02×10^{23} molecules of NO.
- 3g of NO (0.1 mole) contains 6.02×10^{22} molecules of NO.
- 6g of NO (0.2 mole) contains 1.20×10^{23} molecules of NO.

Concentration Calculations (Solution Solutions!)

There are many ways we can express the concentration of one substance dissolved in another, but in ECOT only three methods are used. They are molarity, percent composition, and parts per million (or billion or trillion).

Molarity

Molarity, which has the symbol *M*, is defined as *moles of solute/liter of solution*. Its mathematical expression is

$$M = \left(\frac{\text{mol}}{\text{L}}\right)$$

To make up 1 L of a 1 *M* solution of sodium chloride, for example, we would weigh out 1 mole (58.5 g) of NaCl, place that in a special flask that is marked for exactly 1 L, and add water until we had 1 L of solution. Have we added 1 L of water? No, because the NaCl takes up some of the volume in the flask. We have added less than 1 L of water, but the volume of the solution is exactly 1 L. This is an important point to remember about molarity.

Look at some example problems using molarity.

Example 4. The typical lab concentration of dilute HCl is 6*M*. a. How many moles of HCl are there in a liter of this solution? b. How many moles are there in 250 mL of this solution? c. What volume of this reagent must be delivered to a reaction requiring 4.75 moles of HCl?

Solution:

a. To solve problems in molarity we use the definition of molarity:

$$M = \left(\frac{\text{mol}}{\text{L}}\right)$$

We know the molarity (6 mol/L), we know the volume, 1 L, so we can calculate the number of moles.

$$1\,\text{L}\left(\frac{6\ \text{mol}}{\text{L}}\right) = 6\ \text{mol of HCl}$$

b. Since volume must be in liters, we must convert mL to L.

$$250\ \text{mL}\left(\frac{1\ \text{L}}{1000\ \text{mL}}\right)\left(\frac{6\ \text{mol}}{\text{L}}\right) = 1.5\ \text{mol of HCl}$$

c. We know the number of moles (4.75 mol) and we know the molarity (6 mol/L).

$$4.75\ \text{mol}\left(\frac{1\ \text{L}}{6\ \text{mol}}\right) = 0.792\ \text{L or } 792\ \text{mL of } 6M\ \text{HCl}$$

Example 5. Concentrated hydrochloric acid is 37.9 % HCl by weight and has a density of 1.18 g/mL. Calculate its molarity.

Solution:

In order to calculate the molarity of this solution we have to know the number of moles of HCl in the solution. To find the number of moles we need to know *how many grams* of HCl are in the solution. Since no volume is given we will assume 1 L of solution.

The density is 1.18 g/mL. There are 1000 mL in 1 L, so the weight of 1 L is 1180 g.

37.9 % of that is HCl. The molecular weight of HCl = 1.0 + 35.5 = 36.5 g/mole.

$$\text{weight of HCl} = 0.379 \times 1180 \text{ g} = 447 \text{ g HCl}$$

$$\text{mol of HCl} = 447 \text{ g} \left(\frac{1 \text{ mol}}{36.5 \text{ g}} \right) = 12.2 \text{ mol}$$

$$M \text{ of the solution} = \left(\frac{12.2 \text{ mol}}{L} \right) = 12.2 \, M$$

One operation a chemist often has to carry out in lab is the *dilution* of a concentrated solution (one containing a large amount of solute/volume of solution) to make a more dilute solution (one containing a smaller amount of solute/volume of solution). Chapter 9 in ECOT begins with a dilution problem, which, of course, we assume you have tried!

Example 6. Your lab class needs a total of 750 mL of 3.25 *M* HCl for the day's experiment. How much 6*M* HCl should the lab assistant use to make this up?

Solution:

The key to a dilution problem is that the number of moles of solute is constant. So if we know the number of moles of solute in the one solution, we can use that number of moles to solve for the volume of the other solution.

We are given the volume and the molarity of the diluted solution, so we can use those quantities to find the moles of that solution. We'll convert mL to L in the calculation.

$$\text{mol of HCl} = 750 \text{ mL} \left(\frac{1 \text{ L}}{1000 \text{ mL}} \right) \left(\frac{3.25 \text{ mol}}{L} \right) = 2.44 \text{ mol}$$

We need to take just enough of the more concentrated solution to contain 2.44 mol of HCl, and dilute it to the required volume.

$$\text{vol of } 6M \text{ HCl} = 2.44 \text{ mol} \left(\frac{1 \text{ L}}{6 \text{ mol}} \right) = 0.407 \text{ L or } 407 \text{ mL}$$

Percent Concentration (Percent by Weight, Weight/Weight %, or w/w%)

$$\% \text{ by weight} = \left(\frac{\text{g of solute}}{\text{100 g of solution}} \right) \times 100$$

The most important thing to remember is that the "100 g of solution" above contains the "g of solute" *and* the grams of solvent (usually water). So when we're talking about percent, just remember that the denominator contains "the whole."

$$\% = \left(\frac{\text{part}}{\text{whole}} \right) \times 100$$

Example 7. How many grams of KF are there in 52 g of a 12% aqueous solution?

Solution:
$0.12 \times 52 \text{ g} = 6.2 \text{ g of KF}$

Example 8. What is the percentage concentration of a solution made up of 10 g of $PtCl_4$ and 145 g of water?

Solution:
$\% = \left(\dfrac{\text{part}}{\text{whole}} \right) \times 100$
$= \left(\dfrac{10 \text{ g } PtCl_4}{10 \text{ g } PtCl_4 + 145 \text{ g } H_2O} \right)$
$= 6.45 \%$

Parts Per Million (or Billion or Trillion)

Just as 1% (1 part per hundred) can be illustrated by 0.01 g (1×10^{-2} g), one part per million (ppm) can be represented by 1×10^{-6} g.

Example 9. How many grams of Na are in 500 mL of water which just meets maximum levels for drinking water? The density of water is 1 g/mL.

Solution:
Refer to Table 9.2 of ECOT which shows that 160 ppm of Na are allowed. Since the density of water is 1 g/mL, 500 mL of water weighs 500 g.
Thus, g of Na = $160 \times 10^{-6} \times 500\text{g} = 0.08 \text{ g}$.

Questions Answered

9.1   a. 52 cards;   b. 9 people;   c. 2 pianists;   d. 1000 grams;   e. 10 years.

9.2   The ratio of the mass of a CO_2 molecule to a C atom is 44 amu/12 amu.

$$36 \text{ g of C} \left(\frac{44}{12} \right) = 132 \text{ g of } CO_2$$

9.3   Each of the compounds has 1 carbon per molecule. If we find the number of moles in 1 g of each of the compounds, we have answered the question, since one mole of any substance has the same number of particles.

$$\text{mol } CO_2 = 1 \text{ g} \left(\frac{1 \text{ mol}}{44 \text{ g}} \right) = 0.023 \text{ mol}$$

$$\text{mol CO} = 1 \text{ g} \left(\frac{1 \text{ mol}}{28 \text{ g}} \right) = 0.036 \text{ mol}$$

Since 1 g of CO is the greater part of a mole, it contains the greater number of C atoms. You could also carry the computation further to obtain the number of molecules directly:

$$1 \text{ g of } CO_2 \left(\frac{1 \text{ mol}}{44 \text{ g}} \right) \left(\frac{6.02 \times 10^{23} \text{ molecules}}{\text{mol}} \right) = 1.4 \times 10^{22} \text{ molecules of } CO_2$$

$$1 \text{ g of CO} \left(\frac{1 \text{ mol}}{28 \text{ g}} \right) \left(\frac{6.02 \times 10^{23} \text{ molecules}}{\text{mol}} \right) = 2.2 \times 10^{22} \text{ molecules of } CO_2$$

9.4   Step 1: Find the balanced equation in this section.

$$CH_4 + 2O_2 \rightarrow CO_2 + 2H_2O$$

Step 2: We have 100 grams of methane, and we want to find the weight of oxygen necessary to react completely.

$$\overset{100 \text{ g}}{CH_4} + \overset{x}{2O_2} \rightarrow CO_2 + 2H_2O$$

Step 3: How many moles of methane do we have?

The molecular weight of methane = 12.0 + 4 (1.0) = 16 g/mole.

$$\text{mol of } CH_4 = 100 \text{ g} \left(\frac{1 \text{ mol}}{16 \text{ g}} \right) = 6.25 \text{ mol}$$

Step 4: The mole ratio of methane to oxygen is 1 to 2.

$$\text{mol of oxygen} = 6.25 \text{ mol } CH_4 \left(\frac{2 \text{ mol } O_2}{1 \text{ mol } C_4H_{10}} \right) = 12.5 \text{ mol}$$

66

Step 5: $\text{grams of } O_2 = 12.5 \text{ mol}\left(\dfrac{32 \text{ g}}{1 \text{ mol}}\right) = 400 \text{ g}$

9.5   Use "tsp" as the abbreviation for "teaspoon." The concentration of NaCl in glass #1 is 1 tsp/glass. In glass #2 the concentration is 0.1 tsp/glass. In glass #3, the concentration is 0.01 (or 1×10^{-2}) tsp/glass. Continuing in the same way, the concentration in glass #7 is 1×10^{-6} tsp/glass.

9.6   a. Molarities of salt and sugar solutions

$$\text{mol wt of NaCl} = 23.0 \text{ g} + 35.4 \text{ g} = 58.4 \text{ g/mol}$$

$$\text{mol NaCl} = 23.4 \text{ g}\left(\dfrac{1 \text{ mol}}{58.4 \text{ g}}\right) = 0.40 \text{ mol}$$

$$M_{NaCl} = \left(\dfrac{0.40 \text{ mol}}{0.400 \text{ L}}\right) = 1.0\,M$$

$$\text{Similarly, mol sugar} = 17.1 \text{ g}\left(\dfrac{1 \text{ mol}}{342 \text{ g}}\right) = 0.0500 \text{ mol}$$

$$M_{sugar} = \left(\dfrac{0.0500 \text{ mol}}{0.400 \text{ L}}\right) = 0.125\,M$$

b.   For salt, there are 6 ten-fold dilutions of solution 1 which is $0.214M$; solution 2, then, is $0.0214M$; solution 3, $2.14 \times 10^{-3}M$; continue this until the concentration at glass 7 is $2 \times 10^{-7}M$. Another way to look at this is to remember that 6 ten-fold dilutions will produce a dilution that is $1 \times 10^{-6} \times$ initial M.

$$\text{final concentration} = \left(\dfrac{1.0\,M}{1 \times 10^{6}}\right) = 1.0 \times 10^{-6}\,M$$

9.7   Don't forget to add the weight of the 2 teaspoons of sugar (the solute) to the weight of the coffee (solvent):

$$\% = \left(\dfrac{\text{part}}{\text{whole}}\right) \times 100$$

$$= \left(\dfrac{2 \times 5 \text{ g}}{(2 \times 5 \text{ g}) + 240 \text{ g}}\right) \times 100$$

$$= 4\%$$

9.8   In other words, which solution has the highest number of moles of molecules/liter of solution, i.e., the highest molarity? Hydrogen peroxide at $0.9M$.

9.9   The levels of Zn are 2-6 mg/100g of food. Our thinking might go like this: 1 mg per 1000 g = 1 ppm; 1 mg per 100g = 10 ppm. 2-6 mg per 100g = 20-60 ppm.

Or using the relationships mathematically:

$$1 \text{ ppm} = \left(\frac{1 \text{ mg}}{1{,}000{,}000 \text{ mg}}\right)\left(\frac{1 \text{ mg}}{1000 \text{ g}}\right)$$

$$\left(\frac{\text{mg}}{\text{g}}\right) \times 1000 = \text{ppm}$$

$$\left(\frac{2 \text{ mg}}{100 \text{ g}}\right) \times 1000 = 20 \text{ ppm}$$

9.10 We can set up a proportion. (Remember: 1000 g = 1 kg.)

$$\text{for } N_2, \quad \left(\frac{0.01 \text{ g}}{10^3 \text{ g}}\right) = \left(\frac{x}{10^6 \text{ g}}\right)$$

$$x = 10 \text{ ppm}$$

So, for O_2, x = 50 ppm

9.11 Lead, mercury, silver.

9.12 If the sodium concentration is greater by 50% or 1.5 times greater, then the concentration of sodium in the third glass is 1.5 x 80 or 120 ppm. This is still below the standard, so pollution still begins in glass 2.

You can prove this to yourself by working through the entire problem using 1.5 x 2 or 3 g of sodium:

$$\frac{3 \text{ g sodium}}{250 \text{ g solution}} = \frac{x}{1000 \text{ g solution}}$$

$$x = 12 \text{ g or } 12000 \text{ mg} = 12000 \text{ ppm}$$

In glass 1, concentration of sodium = 12,000 ppm

glass 2, conc = 1200 ppm

glass 3, conc = 120 ppm

Since the Federal Standard is 160 ppm of sodium, glass 2 is still the glass where pollution begins.

Supplementary Exercises

S1. Balance the following equations by completing the equation with the missing reactant or product where necessary, and by placing the proper coefficient in the space provided.

a. _____ H_2O_2 → _____ H_2O + _____ O_2

b. _____ Cs + _____ → _____ CsCl

c. _____ Al + _____ → _____ Al_2O_3

d. _____ H_2SO_4 + _____ NaOH → _____ Na_2SO_4 + _____ H_2O

e. _____ $CuSO_4$ + _____ → _____ $Cu(OH)_2$ + _____ Na_2SO_4

f. _____ C_3H_8 + _____ O_2 → _____ CO_2 + _____

g. _____ C_8H_{18} + _____ → _____ CO_2 + _____ H_2O

S2. Methane, CH_4, the main component of natural gas burns to form water and poisonous carbon monoxide, CO, when the air supply is restricted. Write the balanced chemical equation for this reaction and calculate how much CO can be produced when only 1000 g of O_2 is available.

S3. Concentrated nitric acid is a 68% solution of HNO_3 in water and has a density of 1.41 g/mL. Calculate the molarity of this solution.

S4. Exactly 5 liters of 6M HNO_3 are needed for a lab experiment. What volume of the solution in Exercise #2 would you dilute to make that concentration?

S5. The allowable level of Pb in the blood of children is 0.00015 g/L; the density of normal blood is 1.05 g/mL. What is the allowable concentration in ppm?

S6. Consider a solution made of 17.5 g of sugar dissolved in 500 g of water. How much sugar must be added to this mixture to bring the concentration of sugar to 5.0%?

Where You Might Goof Up...

- Neglecting to have a good experience by skipping the demonstration recommended at the beginning of ECOT's Chapter 9.

- Forgetting to include *all* components in the denominator when calculating percentage composition.

- Forgetting that the mole is just a unit of items, like the dozen.

- Not working all the problems involving concentration and dilution before the quiz.

- Forgetting that the first step in working a stoichiometry problem is a balanced chemical equation.

- Not working all the stoichiometry problems in ECOT and in this *Guide* for practice.

- Not having developed the skill to work with ppm and ppt.

- Being unprepared to argue the thesis "Nothing is absolutely pure."

CHAPTER 10

Acids and Bases

IF IT TASTES SOUR IT MUST BE AN ACID

Jamal Turner looked ruefully at the computer room he shared with two other engineers in the firm they had founded together in the old Wayne building. A fine white dust covered everything, even the partly-eaten sandwich he had left yesterday. What a mess, he thought. As the others arrived each commented on the problem.

"What is this stuff?"

"Is it safe for us to be in here?"

"Where did this come from."

Jamal looked thoughtful. "You know, there was a problem at the lime kiln down the street yesterday. Dust went everywhere. I bet last night's wind blew this in. This could be calcium oxide."

"It could be. The old building's not very tight," said Angela Ballard, "but it could also be something nasty from the ventilation ducts. How are we going to find out? None of us is a chemist, you know."

Jimmy Ho grinned. "I hope we're not depending on Jamal to figure this out. Remember what grade he got in chemistry." They all laughed.

Jamal feigned hurt. "But I did learn something," he pouted. "As a matter of fact, I happen to know how to find out something about this stuff. Let's find a broom, a dustpan, a glass of water, and some grape juice."

"Grape juice!" Jimmy and Angela laughed. "What are you going to do, Jamal, take some refreshment?"

"I'm serious. Get the other stuff, and I'll get some grape juice out of the vending machine."

They returned with everything Jamal had requested. He swept up a small pile of powder and then added a few drops of water to make a white slurry. Then he added a small amount of purple grape juice. The slurry turned green.

"Wow!" said Angela.

"That's impressive," said Jimmy.

"Grape juice turns green in base, red in acid," Jamal said. "Elementary, my dear friends!"

"OK," said Jimmy. "It's a base, it's partially soluble in water, so it's probably an inorganic base which contains hydroxide ion, or it could be a metal oxide, like calcium oxide, that reacts with water to produce hydroxide."

"Right," agreed Jamal. "But I'm not sure how to figure out if it contains calcium ion."

"Ah!" exclaimed Angela. "I know how to do that!"

To be continued in Chapter 13 . . .

Chapter Overview

• A demonstration with cabbage extract is used to distinguish between acids and bases. Practice it at home yourself before amazing your friends.

• The litmus test is described.

• Acids, bases and salts are defined.

• Arrhenius's definition of acids is offered.

• The Brønsted-Lowry theory, a more inclusive definition of acids and bases, is found to be necessary.

- Why cabbage extract behaves the way it does in the presence of acids and bases is explained.

- It is shown that pure water is at the same time both acidic and basic, and therefore neutral.

- The idea of equilibrium as a condition in pure water is introduced.

- pH, the yardstick for measuring $[H^+]$, is defined.

- The concept of relative strengths of acids and bases is introduced and some calculations are performed.

- The names, structures, and activities of some interesting acids are given.

- The effects of acids and bases on us is discussed.

- Antacids are characterized.

- Le Châtelier's Principle, which governs the behavior of chemical equilibria, is described and its influence is demonstrated.

- How blood maintains a fairly constant pH is explained.

For Emphasis

Arrhenius Acids and Bases

This is a good place to go through the text and organize the properties of acids and bases into an information table like we did in chapter 4. Set it up in such as way that you can easily distinguish between these two important classes of chemicals.

One way we know whether a solution is acidic or basic is to use an *indicator*. Some common indicators mentioned by ECOT are litmus, phenolphthalein, cabbage extract, and tea. Add to this the grape juice used by the characters in the introductory scenario above.

It's a good idea to drill yourself on writing neutralization equations. They are a very important part of this chapter. A few examples are below. Write some others.

$$NaOH + HCl \rightarrow NaCl + H_2O$$
$$KOH + HBr \rightarrow KBr + H_2O$$
$$LiOH + HF \rightarrow LiF + H_2O$$
$$2\ KOH + H_2SO_4 \rightarrow K_2SO_4 + 2\ H_2O$$
$$Ca(OH)_2 + 2\ HBr \rightarrow CaBr_2 + 2\ H_2O$$
$$NaOH + CH_3COOH \rightarrow CH_3COONa + H_2O$$

Brønsted-Lowry Acids and Bases

We've seen that acids and bases give hydrogen ions and hydroxide ions to make salts and water, but if you did the demonstration with the cabbage-leaf extract you found out that ammonia is also a basic solution. Ammonia, NH_3, has no hydroxide in its formula. Yet an indicator shows that it's solution is basic. How can ammonia be a base? Where does the OH^- come from? Since the NH_3 gives a basic reaction, the list of acid and base properties in ECOT is apparently inadequate. Arrhenius's ideas of what an acid and a base are must be less than all-inclusive. Obviously, a bigger definition of acids and bases is needed. But before we look for that, let's consider how ammonia causes a basic reaction. An ammonia solution is really a solution of NH_3 gas in water.

$$NH_3 + H_2O \rightarrow NH_4^+ + OH^-$$

The reaction above describes a process whereby the NH_3 extracts a proton, H^+, from the water to become an ammonium ion, leaving a hydroxide ion as the other product. Instead of giving an OH^- ion to solution, it takes a proton from the solvent. Also, we see that the water *donates* a proton, a hydrogen ion, to the ammonia! Doesn't this make the water an acid? Isn't water thought to be neutral? What are we going to do about this?

We're going to change our thinking about acids and bases, that's what. Our new definitions must include everything in the old set above plus enough new ideas to account for the ammonia reaction above; it would be handy if it covered other bases like ammonia and other "acids" like water. As noted in ECOT, Brønsted and Lowry proposed a broadening of our way of thinking about acids and bases. Not only do their new definitions provide for ammonia and water, but for other situations not covered by Arrhenius. Here are their ideas:

- Acids *donate* protons. This is not too different from the Arrhenius definition, but there are inferences to be made beyond this simple statement.

- Bases *accept* protons. This is quite different from the old definition in which an OH^- had to be present in the compound to begin with.

- A compound is either acidic or basic *only with respect to something else.*

- An acid is *strong* if it gives up a proton easily. A base is *strong* if it has a powerful attraction for a proton.

Metal Oxides and Nonmetal Oxides

Metals and nonmetals play a role in the study of acid-base chemistry. Some examples:

metal oxide + water $\rightarrow$ base
$Na_2O + H_2O \rightarrow 2\,NaOH$
$BaO + H_2O \rightarrow Ba(OH)_2$

nonmetal oxide + water $\rightarrow$ acid
$CO_2 + H_2O \rightarrow H_2CO_3$
$SO_2 + H_2O \rightarrow H_2SO_3$

A metal oxide is the subject of the introductory scenario (see Supplementary Exercise S1). And you have seen the reaction of CO_2 with water several times in ECOT. In the last reaction above, SO_2, a major player in air pollution, reacts to form sulfurous acid, one of the constituents of acid rain you will study in greater detail in Chapter 14.

pH — A Short-Hand Way of Treating Acidity

The concept of pH is one of the handiest ideas chemists have developed. We are all in Sørensen's debt. It makes very small numbers easy to handle and allows for quick thinking in clinical situations. Which is easier: "Doctor, your patient's blood pH is down to 7.05; should we do something?" or "Doctor, your patient's blood's hydrogen-ion concentration is down to 8.91×10^{-8}; should we do something?" The discussion in Section 9.6 of ECOT presents the equilibrium that exists in *all* aqueous solutions. Whatever has been dissolved in the water—battery acid, window cleaner, chicken soup, tea, plant food, Kool-Aid or white-side-wall tire cleaner— the equilibrium and its constant holds true:

$$2H_2O \leftrightarrows H_3O^+ + OH^- \quad K_w = [H_3O^+][OH^-] = 1 \times 10^{-14}$$

The implication of the relationships above is that when the $[H_3O^+]$ goes up, the $[OH^-]$ goes down, and *vice versa*. *In a neutral solution only*, $[H_3O^+] = [OH^-] = 1 \times 10^{-7}$, and the pH = 7. Also, the pOH = 7 in neutral solution. The change in the pOH with changing pH is what you would expect as shown by the table below.

$[H_3O^+]$	pH	pOH	$[OH^-]$
1	0	14	1×10^{-14}
1×10^{-5}	5	9	1×10^{-9}
1×10^{-10}	10	4	1×10^{-4}
1×10^{-14}	14	0	1

Notice the special relationship among the data in the table: pH + pOH = 14. Remember:

$$K_W = [H_3O^+][OH^-] = 1.0 \times 10^{-14} \qquad [H_3O^+] = K_w/[OH^-] \text{ and } [OH^-] = K_w/[H_3O^+]$$

Questions Answered

10.1   a. Blue; b. red.

10.2 $H_2SO_4 + 2\,KOH \rightarrow K_2SO_4 + 2\,H_2O$
       $2\,HI + Ca(OH)_2 \rightarrow CaI_2 + 2\,H_2O$
       $H_2CO_3 + Mg(OH)_2 \rightarrow MgCO_3\ 2\,H_2O$

10.3   a. Yes; b. no. An Arrhenius base is a compound that produces OH⁻ in water, and this can occur if an acid is not present. A Brønsted-Lowry base accepts a proton from an acid. If no acid is present to donate a proton, then a Brønsted-Lowry base cannot be present.

10.4 $HCl + H_2O \rightarrow Cl^- + H_3O^+$

10.5 1×10^{-3} M; 0.00001 M

10.6   a. 0.01 M HCl; b. 0.01 M acetic acid; c. 0.01 M HCl; d. 0.01 M HCl.

10.7 See Table 10.3 of ECOT.
       a. 1×10^{-11};

b. If the pH of ammonia is 11: $K_w = [H^+] [OH^-] = 1 \times 10^{-14}$; $[H^+] = 1 \times 10^{-11}$. So the $[OH^-] = 1 \times 10^{-3}$.

c. 3.

10.8   a. Butyric acid; b. oxalic acid; c. lactic acid; d. propionic acid; e. benzoic acid; f. lactic acid; g. citric acid; h. formic; i. acetic acid.

10.9   Normal stomach activity would cease; acid rebound could occur.

10.10 Using two equations in this section, we can write the equilibrium of carbonic acid in the blood as

$$CO_2 + H_2O \rightleftharpoons H^+ + HCO_3^-.$$

Hyperventilation removes CO_2 from the body, thus depleting CO_2 on the left side of the equation above. To readjust, the equilibrium must shift to the left, producing more CO_2 and H_2O and reducing H^+. A reduced concentration of H^+ means a higher pH.

Supplementary Exercises

S1.   a. Write the equation for the reaction between calcium oxide and water that Jamal performed in the introductory scenario.

b. What would you expect MgO to do in the same situation? Write the equation.

S2.   Complete and balance the following acid-base reactions:

a)   KOH + HBr →

b)   KOH + H_2SO_3 →

c) $Ca(OH)_2$ + HF →

d) $Sr(OH)_2$ + H_3PO_4 →

S3. Calculate the following:

    a. pH of $0.001M$ HCl;

    b. pH of $0.001M$ NaOH;

    c. pOH of $10^{-2}M$ HCl

S4. Which has the greater $[H_3O^+]$?

    a. A solution with pH = 1 or a solution with pH = 2;

    b. A solution with pH = 10 or a solution with pH = 11;

    c. A solution with pOH = 3 or a solution with pOH = 5;

    d. A solution with pOH = 10 or a solution with pOH = 12.

S5. How much $0.2M$ HCl is necessary to neutralize each of the following?

    a.   250 mL of 0.1 M $Ca(OH)_2$;

    b.   3.7 g of CaO;

    c.   1 L of 0.1 M CH_3COOH.

Where You Might Goof Up...

- Neglecting to put signs on ions where appropriate.

- Forgetting the few simple formulas needed for calculating concentrations of acids and bases.

- Being unable to identify an acid as weak or strong.

- Being unable to predict the effects of change in concentration on an equilibrium according to Le Châtelier's principle.

- Being unable to describe qualitatively and quantitatively the connection between pH and $[H^+]$.

- In a discussion of chemical equilibrium, omitting comments about its dynamic aspects.

- Being unable to distinguish between Arrhenius and Brønsted-Lowry acid-base systems.

- Being unable to list a dozen or so acids and bases that are common to everyday life.

- Being unable to use K_w to calculate $[H_3O^+]$ given $[OH^-]$ and *vice versa*.

- Being unable to describe the effect of an antacid on stomach chemistry.

CHAPTER 11

Oxidation and Reduction

THE ELECTRICITY OF CHEMISTRY

"Come on, Marta. We're going to be late."

"Just a minute, Joe. I have to change the zinc electrodes in the car battery. Did you get the extra electrodes? At 250 miles/set, we'll need at least 4 electrode changes for the round trip."

"I got 'em. Let's go. " Joe got in the car, and as Marta completed the electrode change, he turned the switch to "On". They heard the quiet purr of the fan that draws air over the zinc electrode and then the car's computer displayed voltage and energy density of the zinc-air battery. The electric motor started, and they headed for the Interstate, merging with 65 mi/hr traffic.

"It doesn't have quite the "zip" our gasoline-powered car had," remarked Joe, "but it certainly has the speed on the highway. I think this was a good buy. "

Marta nodded. "I'm especially pleased that the zinc electrodes are cheap and the power is clean." She grinned. "Let me know when it's my turn to drive!"

Although fully electrically-powered vehicles for high speed and long-distance travel are not yet commercially viable, hybrid electric vehicles (HEVs) are on the market in 2002. The Honda Insight and Toyota Prius are two of the available HEVs. Both use a nickel metal hydride (Ni-MH) battery similar to those used in many computers. The disadvantages of the Ni-MH batteries include their relatively high cost and lack of recyclability. The lead acid battery has potential for HEV application, but its disadvantages (weight, poor performance at cold temperatures and short life) outweigh the advantage of its extremely low cost. The zinc-air battery is only one of several new technologies to arrive in the last decade on the battery market. Whether it will find its way into the HEV market remains to be seen. As our requirements for clean, cheap power increase we can expect to see major changes in everything from children's toys (batteries still not included?), to computers, to automobiles.

Chapter Overview

• A colorful experiment that can be done at home makes visual the phenomena of oxidation and reduction.

• A flashlight battery's innards are described.

• The construction of the basic potato cell is outlined.

• The Daniell cell is discussed, and the electrochemistry of Zn and Cu is developed.

• Reduction and oxidation (REDOX) are defined and the idea of a half-cell reaction is presented.

• Voltage and amperage are dealt with.

• The hydrogen electrode, a reference point, is characterized.

• Elements and ions are ranked in their tendency to oxidize or reduce, and standard reduction potentials are introduced.

- You are shown how to predict whether a certain redox reaction will produce electricity spontaneously; that is to say, "Will this battery work?"

- The battery demonstration from Chapter 1 of ECOT is revisited, this time with a fuller explanation of "why."

- The innards of various batteries, including the automobile battery, are explored.

- An important industrial process, electrolysis, is pictured, and applications are enumerated.

- Applications of electrochemistry to swimming pools, rust, and electric cars are listed.

For Emphasis

Breaking Redox into Half-cells

If one species gains electrons, the other one has to lose them; if one species is reduced, the other has to be oxidized. A half-cell reaction can represent either an oxidation or a reduction, but for two half-cell reactions added together to have any meaning, one of them has to be an oxidation half-cell reaction and the other has to be a reduction half-cell reaction. We can see how this is so by looking at what happens when we add two half-cell reactions. In Section 11.5 of ECOT, we see how to add two half-cell reactions together, particularly the Zn and Cu half-cells where Zn is being oxidized by Cu^{2+}. Let's look at an electrochemical cell in which Cu^{2+} is reduced to Cu and Fe is oxidized to Fe^{2+}:

$$\text{Reduction:} \quad Cu^{2+} + 2e \rightarrow Cu^0$$
$$\text{Oxidation:} \quad Fe^0 \rightarrow Fe^{2+} + 2e$$

$$\text{Redox:} \quad Cu^{2+} + Fe^0 \rightarrow Cu^0 + Fe^{2+} \quad\quad (1)$$

Notice that the two electrons cancel; we are left only with chemical species, either elements or ions. If by accident we write the equations so that the sum includes electrons, we know right away that we have made an error.

If the number of electrons in the two half-cell reactions are not equal, one of the half-cell reactions must be rewritten; let's see how this is accomplished. Consider the reaction between sodium metal and chlorine gas to form sodium chloride. We know from previous discussions that Na is oxidized and chlorine gas is reduced in this reaction. We can write the half-cells for each of them as follows.

$$\text{Oxidation:} \quad Na \rightarrow Na^+ + e$$
$$\text{Reduction:} \quad Cl_2 + 2e \rightarrow 2\,Cl^-$$

Plainly, we cannot add these two half-cells together as they stand and eliminate all the electrons. However, we can solve this problem by multiplying the Na half-cell through by two. Now we are able to deal with the system in the same manner as before.

$$\text{Oxidation:} \quad 2\,Na \rightarrow 2\,Na^+ + 2e$$
$$\text{Reduction:} \quad Cl_2 + 2e \rightarrow 2\,Cl^-$$

$$\text{Redox:} \quad 2\,Na + Cl_2 \rightarrow 2\,Na^+ + 2\,Cl^- \quad\quad (2)$$

After we have multiplied the Na half-cell equation by two, we can add it to the Cl_2 half-cell and the electrons cancel out. Notice also, that this results automatically in a *balanced* equation. Occasionally it will be necessary to multiply both equations by different integers to eliminate the electrons.

Will the Reaction "Go"?

Beyond saying that some compound is a "good" reducing agent or a "weak" oxidizing agent, we can quantify the tendency of reactants to produce products; in other words, we can say whether a reaction will occur *spontaneously*. Table 11.1 of ECOT shows the voltages obtained when each of the half-cells is measured against the hydrogen electrode. Locating the Cu half-cell, we see that the tendency for Cu^{2+} to be reduced is +0.34 v compared to the hydrogen electrode.

This table is most useful since the higher the voltage (including the sign), the greater the tendency to become reduced. You have been adding two half-cell reactions together to get a full chemical reaction or description of the chemical process going on in a battery. You can also add the voltages of the half-cells together to decide whether a reaction will "go" (occurs spontaneously) or not. This is really an easier way to decide if a reaction is feasible than going to the lab. There is only one way to get in trouble with this: *forgetting to keep track of the signs of the voltages*. For instance, when we write the Cu half-cell above as an *oxidation*; it is the reverse of what is tabulated in the standard reduction potentials list showing +0.34v. If we reverse the half-cell, *we change the sign* of the half-cell potential, in the case as written above, to -0.34v. Writing the half-cells above again with their voltages:

$$\text{Oxidation} \quad Cu^0 \rightarrow Cu^{2+} + 2e \qquad -0.34v$$
$$\text{Reduction:} \quad 2 H^+ + 2e \rightarrow H_2 \qquad 0.00v$$

$$\text{Redox:} \qquad 2 H^+ + Cu^0 \rightarrow H_2 + Cu^{2+} \qquad -0.34v \qquad (3)$$

Since the voltage of *the cell* is less than zero, this reaction will *not* occur spontaneously. This is a general rule: Chemical reactions represented by electrochemical cells will proceed spontaneously only if the cell voltage is positive. There's another general rule which allows us to predict spontaneity without having to calculate the voltage.

> A species to the *left* of the arrow will *oxidize* any species to the *right* of the arrow *above* it in the table.

So, let's apply this rule to the reaction we just examined between $H^+ + Cu^0$. Find the half-reactions in Table 11.1. They are in this order in the Table:

$$2 H^+ + 2e \rightarrow H_2$$
$$Cu^{2+} + 2e \rightarrow Cu$$

According to the rule in the box above, the species to the left of the arrow (Cu^{2+}) will oxidize any species to the right of the arrow (H_2) above it in the table. In other words, Cu^{2+} will be reduced and H_2 will be oxidized. Is this the reaction we are examining? No. In reaction (3) above, we see the reverse of this. The reaction will go if Cu^{2+} is reduced and H_2 is oxidized. The reaction won't go if Cu^0 is oxidized and H^+ is reduced.

Let's look at Table 11.1 in ECOT, Standard Reduction Potentials, and solve some more problems.

Example: Will a reaction occur spontaneously between aluminum, Al, and liquid bromine, Br_2? If so, what is the voltage of the resulting electrochemical cell?

Solution:

Will it occur spontaneously? Find the two possible reactions in Table 11.1. In the order in which they are in the table they look like this:

$$Al^{3+} + 3\,e \rightarrow Al$$
$$Br_2 + 2\,e \rightarrow 2\,Br^-$$

The species to the left of the arrow will oxidize the species to the right of the arrow above it. So, Br_2 will oxidize Al, which means that the answer to the question is "yes."

What is the voltage of the resulting cell? Rewrite the equations to reflect the reaction which is occurring, and write the standard potentials, being careful to reverse the sign of the one which is oxidized. As in equation (2) above, be sure that the electrons cancel.

$2\,(Al \rightarrow Al^{3+} + 3\,e) =$	$2\,Al \rightarrow 2\,Al^{3+} + 6\,e$	$+1.66$ v
$3\,(Br_2 + 2\,e \rightarrow 2\,Br^-) =$	$3\,Br_2 + 6\,e \rightarrow 6\,Br^-$	$\underline{+1.07}$ v
Sum:	$2\,Al + 3\,Br_2 \rightarrow 2\,Al^{3+} + 6\,Br^-$	$+2.73$ v

We multiplied the Br electrode (half-cell) by 3, why not multiply the voltage by 3? Doesn't the potential become 3.21 v? No; here's why: Voltage, or electrical pressure, is an *intensive* quantity, independent of mass. Temperature is another intensive property. If we have a 2-liter beaker containing one liter of water whose temperature is 25 °C, does the temperature become 50 °C if we add 1 more liter of water at 25 °C? Of course not. Whether we put three or seven gallons of gas in the tank, the temperature of the tank's contents does not depend on how full the tank is. Intensive quantities do not depend on the amount of material present. Whether we multiply a half-cell by 3 or 0.5, the cell potential stays the same.

The array in Table 11.1 of ECOT contains a lot of information, easily accessed. If we restrict ourselves to the half-cells themselves (ignore the voltage column), we can make the following generalizations:

- The best oxidizing agent in the list is on the lower left of the arrow (F_2, in this case).

- The best reducing agent in the list is on the upper right of the arrow (Li, in this case).

- Any species on the left of the arrow will oxidize anything to the right of the arrow above it (Ag^+ will oxidize Ni, for instance).

- The worst reducing agent in the list is on the lower right of the arrow (F^-, in this case).

- The worst oxidizing agent in the list is on the upper left of the arrow (Li^+, in this case).

- A reducing agent will reduce any oxidized species below it. For instance, Cd will reduce Ni^{2+}, I_2 or F_2, but not Na^+.

- An oxidizing agent will oxidize any reduced species above it. For instance, Br_2 will oxidize Ag, Fe or K, but not Cl^-.

82

Electrolysis

Electrolysis can be thought of as the opposite of electrochemical cell processes. Electrochemical cells consume chemicals and produce electricity; electrolysis consumes electricity to produce chemicals. Silver plate flatware is less expensive than pure silver and is made by immersing a copper item in a solution of silver nitrate with silver as the other electrode. With the copper item as the cathode, a direct current power supply is attached across the electrodes. The half-cells for this process can be written:

$$\text{At the cathode:} \quad Ag^+ + e \rightarrow Ag$$
$$\text{At the anode:} \quad Ag \rightarrow Ag^+ + e$$

Similar reactions can be used to plate chromium on bumpers, copper on tin and for other industrial as well as commercial uses. The most common example of the difference between the two kinds of cells (voltaic and electrolytic) occurs during the operation of the automobile: When starting the car, the battery produces the electrical energy to turn the engine over, consuming Pb, H_2SO_4 and PbO_2 and producing $PbSO_4$; when driving down the road, the engine generates electricity and recharges the battery (regenerates the Pb, the H_2SO_4 and PbO_2 from the $PbSO_4$), an electrolytic process.

Questions Answered

11.1 Zinc; ammonium chloride, zinc chloride, manganese dioxide; carbon.

11.2 No. A salt bridge allows ions to move from one part of the cell to the other. Ions can't move through copper wire, and without ion movement the cell doesn't function.

11.3 Since zinc sulfate solution has no color, neither zinc nor sulfate has color; if copper sulfate is blue, it must be due to the copper ions.

11.4 Zinc sulfate.

11.5 Calcium is being oxidized, chlorine is being reduced.

$$Ca \rightarrow Ca^{2+} + 2e$$
$$Cl_2 + 2e \rightarrow 2Cl^-$$

11.6 Yes. The ions would move under the influence of the voltage of the cell. The Cl^- moves toward the Zn half-cell (toward the positive electrode), the Na^+ moves toward the Cu half-cell (toward the negative electrode).

11.7 Increasing the size of the plates does not affect the electrical pressure, voltage, since voltage is an intensive property and depends only on which ions are being oxidized and reduced. Increasing the size of the plates allows for a greater flow of electrons (amperage), and hence more current to be drawn from the cell.

11.8 F_2; Li^+.

11.9 At the top of the table, Li^+ has the most negative reduction potential, indicating that it has the least potential to be reduced. Turn the equation around, and Li has the most positive oxidation potential (+3.04 v) in the table and the greatest tendency to be oxidized. Since reducing agents are themselves oxidized, Li must be the strongest reducing agent.

11.10 To be spontaneous, the sum of the potentials of the oxidation and reduction half reactions must be positive.

Before doing any arithmetic, look at the position of each reduction in Table 11.1 in ECOT. Remember, the species to the left of the arrow will oxidize the species to the right of the arrow above it. H^+ is to the left of the arrow, and it oxidizes the species to the right of the arrow above it. Fe and Zn are to the right of the arrow.

	Oxidation:	$Fe \rightarrow Fe^{2+} + 2e$	+0.44 v
	Reduction	$H_2 + 2e \rightarrow 2 H^+$	+0.00 v
	Sum:	$Fe + H_2 \rightarrow Fe^{2+} + 2 H^+$	+0.44 v

Similarly,

	Oxidation:	$Zn \rightarrow Zn^{2+} + 2e$	+0.76 v
	Reduction	$H_2 + 2e \rightarrow 2 H^+$	+0.00 v
	Sum:	$Zn + H_2 \rightarrow Zn^{2+} + 2 H^+$	+0.76 v

There are quite a few other metals above the hydrogen half reaction. Three are Li, Ca, and Al. Three metals that will *not* liberate hydrogen are below the hydrogen half reaction: Cu, Ag, and Au.

11.11 To decolorize the tincture means to reduce the iodine to iodide. Iodine will oxidize anything above it to the right of the arrow in Table 11.1. Thus it will react with all of the metals except Ag.

11.12 The two extremes in the table are the reductions of F_2 and Li^+. We know that fluorine will be reduced when lithium is oxidized, so

	Oxidation:	$2(Li \rightarrow Li^+ + e)$	+3.04 v
	Reduction	$F_2 + 2e \rightarrow 2 F^-$	+2.87v
	Sum:	$2Li + I_2 \rightarrow 2Li^+ + 2 I^-$	+5.91 v

11.13 No. It must be kept upright at all times, and it would be a trifle messy.

11.14 When the battery is recharged, sulfuric acid is produced, which is denser than water; thus the density increases.

11.15 H_2, Cl_2, and NaOH. In the second case, H_2, O_2, and NaOH. (O_2 is between Br_2 and Cl_2 in Table 11.1.)

11.16 Galvanized metal has its surface coated with Zn. The Zn acts as cladding, protecting the underlying metal from corrosion; it also protects it by being the target of choice for the attacking agent since it is more easily oxidized.

Supplementary Exercises

S1. A strip of lead, Pb, slowly dissolves in hydrochloric acid. Bubbles form.

    a. Write the half reactions and the equation for the overall reaction.

    b. Would you place the Pb half reaction above or below the hydrogen half reaction in Table 11.1 in ECOT (Standard Reduction Potentials)? Explain your answer.

S2. The gold half-cell, $Au^{3+} + 3e \rightarrow Au$, has a reduction potential of 1.50v.

    a. Will gold dissolve in acid?

    b. If a piece of gold is placed in a solution of $FeSO_4$, what will be observed?

S3. Given the following list of standard reduction potentials, answer the questions below.

$$Ag^{2+} + e \rightarrow Ag^+ \qquad +1.98v$$
$$Hg^{2+} + 2e \rightarrow Hg \qquad +0.85v$$
$$Sr^{2+} + 2e \rightarrow Sr \qquad -2.89v$$

(Hint: For the questions below, it may be helpful to rearrange these in the order they would have on the standard reduction potential table.)

    a. What is the strongest oxidizing agent in this list?

    b. What is the strongest reducing agent in this list?

    c. Will a reaction between Sr^{2+} and Hg be spontaneous? Give a reason for your answer.

S4. Add the Cd and H_2O half-cells together in such a way that the reaction "goes."

S5. Add the O_2 and Ni half-cells together in such a way that the reaction "goes."

S6. An electroplating solution containing what ion should be used to coat steel with cadmium?

Where You Might Goof Up...

- Being unable to list the half-cells involved in the nail/iodine demonstration described at the beginning of the chapter.

- Not realizing that two different electrodes placed in any ionic solution will generate a voltage.

- Thinking that the magnitude of the voltage depends on the shape, size or condition (porous, mossy or smooth) of the electrode.

- Forgetting that *O*xidation takes place at the *A*node, and *R*eduction takes place at the *C*athode.

- Forgetting OIL RIG.

- Being unable to distinguish between electrochemical (voltaic) cells and electrolysis cells.

- Being able to list only two examples of voltaic and electrolysis cells each.

- Forgetting to change the sign of a reduction potential when reversing the expression for a reduction half-cell.

- Overlooking the fact that for every reduction, there must also be an oxidation.

- Not explaining the use of a salt bridge adequately.

- Leaving charges off ions, either in text or equations.

- Trying to add half-cells when the electrons do not cancel.

- Misinterpreting the sign of a redox (sum of both half-cells) voltage with regard to whether a reaction will proceed spontaneously.

- Given a list of half-cells with voltages, failing to select the best oxidizing and reducing agent from that list.

- Giving insufficient justification for setting the hydrogen electrode voltage at 0 v.

- Allowing the subtle difference between rust and corrosion to evade understanding.

CHAPTER 12

Solids, Liquids, and Especially Gases

THE STATES OF MATTER

The August day was hot and humid. Paul Berneau and his wife, Bernadette, were working the hay field, hoping to get hay cut and cured before the next rain. Paul's brother, Jean Pierre, was swinging the hay scythe, and Jean Pierre's wife, Marie, was raking hay behind him. Marie lay down her rake and stretched her tired muscles. As she looked up her attention was caught by an object that appeared to be descending from the heavens. Marie gasped. She knew that only terrible, monstrous things were expelled from the heavens, and one was coming toward them. She stood, transfixed, her mouth open, her eyes wide. Jean Pierre, sensing her stillness, turned around.

"Marie!"

Marie couldn't respond. She pointed, and Jean Pierre looked up.

"Paul! Bernadette!" he screamed, pointing at the intruder. Instinctively, haying tools became defensive weapons. The monster came closer and closer, looming larger and larger, until it hit the ground, bouncing threateningly.

"Now!" yelled Jean Pierre, and lunged with the hay fork. He pierced the monster's side, and with a "pooh" gases escaped, and the terrible thing began deflating. Encouraged by the decreasing size and apparent suffering of the monster, the others hit it with their tools, until finally the intruder lay in ruins.

The farmers looked at each other, daring finally to take a breath. In the stillness they heard a galloping horse, and they looked up to see a rider careering toward them shaking his fist.

"What have you done? What have you done?" shouted the rider. He dismounted in a swirl of dust and examined the scene. He shook his head.

"You have destroyed Jacques Charles's hydrogen balloon, the first such in the whole world," he lamented, as he stared at the shreds of rubberized cloth and rope.

Thus ended the first experiment with a hydrogen balloon, following only a few months behind Joseph Montgolfier's first hot air balloon. Before the end of 1783 both types of balloons were to carry humans aloft.

Early experiments in hot air and hydrogen ballooning, seeming crude to us, perhaps, in the early 21st century were every bit as remarkable in their time as the flight to the Moon. Space flight has given us not only a greater understanding of space, but it has revolutionized the way we view the Earth. The first view from space of the beautiful, fragile blue ball that is the Earth gave a big push to an incipient environmental movement. In the same way, the experiments of intrepid balloonists 200 years ago made atmospheric studies possible for the first time and dispelled many superstitions. Meteorological studies began as scientists like Gay-Lussac ascended to 23,000 feet to study the weather. Balloons are still used today to study weather and high altitude air flow patterns, for obtaining air samples in pollution studies, even for observing stars from a better vantage point than the surface of the Earth provides.

Chapter Overview

- A demonstration is described wherein a gas law (Charles' Law) can make you look very strong indeed.

- How solids, liquids and gases differ is explained.

- Melting and boiling are described at the molecular level.

- The regions of our planetary gaseous envelope are described, and atmospheric pressure is defined.

- The regions of our planetary gaseous envelope are described, and atmospheric pressure is defined.

- An ideal gas is defined, and the kinetic-molecular theory is outlined.

- Boyle's and Charles' laws are introduced and some calculations are performed.

- Gay-Lussac's and Avogadro's laws describing the chemical reactions of gases are detailed.

- How the gas laws (especially Henry's and Dalton's) control the behavior of soda pop is demonstrated.

- Our respiratory system is described, and how it works is related.

For Emphasis
Boyle's, Charles', and the Combined Gas Law

There are several ways that people generally approach the gas laws, particularly Boyle's and Charles' laws and the combined law which brings the two together. You may be more comfortable memorizing formulas for solving gas law problems, or you may enjoy the challenge of thinking through each problem so that you don't have to memorize formulas. These techniques are illustrated below. Choose the method that works for you, and apply it to the problems at the end of Chapter 12 in ECOT and to the Supplementary Problems below. Remember, however, that no matter what method you choose to apply to solving these problems, you have to know the gas laws so well that you recognize immediately whether it is a Charles' Law or a Boyle's Law problem or a combination of the two.

Let's work an example by several methods.

Example 1: A sample of gas occupies a volume of 100 mL at 750 mm Hg. If temperature is held constant, what volume will the gas occupy at 320 mm Hg?

Recognize this as a Boyle's Law problem. The key is that pressure and volume are changing and temperature is constant. No matter what method of solution you choose, first make a table of what's given and what's to find:

Initial volume: $V_1 = 100$ mL
Final volume: $V_2 = ?$
Initial pressure: $P_1 = 750$ mm Hg
Final pressure: $P_2 = 320$ mm Hg

Solution method 1:

The author of ECOT uses $P_1 \times V_1 = k$ to solve Boyle's Law problems.

$$P_1 \times V_1 = k$$
$$\text{and} \quad P_2 \times V_2 = k$$
$$\text{So,} \quad P_1 \times V_1 = P_2 \times V_2$$
$$\text{solving for } V_2, \quad V_2 = \frac{P_1 \times V_1}{P_2}$$
$$= \frac{750 \text{ mm Hg} \times 100 \text{ mL}}{320 \text{ mm Hg}}$$
$$= 234 \text{ mL}$$

Solution method 2:

Memorize one expression of Boyle's Law, such as $P_1V_1 = P_2V_2$. Solve it for the unknown.

$$P_1 V_1 = P_2 V_2$$

$$\text{solving for } V_2, \quad V_2 = \frac{P_1 \times V_1}{P_2}$$

$$= \frac{750 \text{ mm Hg} \times 100 \text{ mL}}{320 \text{ mm Hg}}$$

$$= 234 \text{ mL}$$

Solution method 3:

This is the preferred method of one of the *Guide's* authors (AR). Forget the formulas. Simply keep in mind what Boyle's law says: if we increase the pressure on a gas, the volume decreases (temperature constant). Or if we decrease the pressure, the volume increases. Use the graphic of Fig. 12.10 in ECOT to fix this in your mind.

If you look at either solution method above you will see that V_2 is equal to V_1 multiplied by a factor of one pressure over the other. We can determine intuitively what that factor is.

$$V_2 = V_1 \left(\text{factor of pressures} \right)$$

What's happening in the problem? Pressure is decreasing (from 750 to 320 mm Hg) so Boyle's law says that volume must increase. Therefore to make V_2 *larger* than V_1, we must multiply V_1 by a factor of pressures which is *greater than* 1. The larger pressure divided by the smaller pressure gives a factor greater than 1.

$$V_2 = 100 \text{ mL} \left(\frac{750 \text{ mm Hg}}{320 \text{ mm Hg}} \right)$$

$$= 234 \text{ mL}$$

Example 2: A gas has a volume of 423 mL at 100 °C. It is cooled to 0 °C, with the pressure held constant. What is the new volume?

Remember, no matter how you approach the problem, make a table of the given information and note what is to be found. Also, recognize that this is a Charles' law problem. Temperature and volume are changing, and pressure is constant. Since there is no mention of a change in the amount of gas we can assume that it is constant. But notice what we have to do to Celsius temperatures: In gas law problems, *always convert Celsius temperatures to Kelvin before you start solving the problem.*

Initial volume: $V_1 = 423$ mL
Final volume: $V_2 = ?$
Initial temperature: $T_1 = 100$ °C + 273 = 373 K
Final temperature: $T_2 = 0$ °C + 273 = 273 K

Solution method 1:

$$V_1 = k \times T_1 \text{ and } V_2 = k \times T_2$$

So, $k = \dfrac{V_1}{T_1}$ and $k = \dfrac{V_2}{T_2}$

therefore, $\dfrac{V_1}{T_1} = \dfrac{V_2}{T_2}$

and $V_2 = ?$

Solve for V_2.

Solution method 2:

Memorize the formula, $\dfrac{V_1}{T_1} = \dfrac{V_2}{T_2}$, and solve for V_2.

Solution method 3:

$$V_2 = V_1 \text{ (factor of temperatures)}$$

Reasoning: Charles' law says that if temperature is decreasing, volume must decrease. V_2 will be less than V_1, so the factor of temperatures must be less than 1.

$$V_2 = 423 \text{ mL} \left(\frac{273 \text{ K}}{373 \text{ K}} \right)$$

$$V_2 = 310 \text{ mL}$$

Now let's combine the two gas laws. The author of ECOT gives us the combined formula derived from Boyle's and Charles' laws. It is

$$\frac{V_1 P_1}{T_1} = \frac{V_2 P_2}{T_2}$$

Memorize this formula and use it, or use solution method 3 to think through each problem.

Example 3: A balloon has a volume of 450 L at 25 °C and 1 atm. What will its volume be at 0.5 atm and -40 °C? Which factor, pressure change or temperature change, is the dominant factor in determining the new volume?

$V_1 = 450 \text{ L}$
$V_2 = ?$
$P_1 = 1 \text{ atm}$
$P_2 = 0.5 \text{ atm}$
$T_1 = 25 \text{ °C} + 273 = 298 \text{ K}$
$T_2 = -40 \text{ °C} + 273 = 233 \text{ K}$

90

Solution:

$$V_2 = V_1 \text{ (factor of pressures) (factor of temperatures)}$$

What's happening? Pressure is decreasing, volume must increase; temperature is decreasing, volume must decrease.

$$V_2 = 450 \text{ L}\left(\frac{1 \text{ atm}}{0.5 \text{ atm}}\right)\left(\frac{233 \text{ K}}{298 \text{ K}}\right)$$

$$= 704 \text{ L}$$

Notice that it's not necessary to convert the pressures from atmospheres to mm Hg. As long as the units are the same, they will cancel.

The dominant factor in determining the new volume is the pressure factor. The temperature factor is just slightly less than 1, so its contribution to the new volume is small.

Example 4: 400 mL of a gas are collected at 37 ˚C and 400 mm Hg. At what Celsius temperature will the volume be 300 mL and the pressure 1.2 atm?

$$V_1 = 400 \text{ mL}$$
$$V_2 = 300 \text{ mL}$$
$$P_1 = 400 \text{ mm Hg atm}$$
$$P_2 = 1.2 \text{ atm} = 1.2 \text{ atm} \times 760 \text{ mm Hg/atm} = 912 \text{ mm Hg}$$
$$T_1 = 37 \text{ ˚C} + 273 = 310 \text{ K}$$
$$T_2 = ? \text{ K} - 273 = ? \text{ C}$$

Solution:

Notice that before we start it's necessary to put pressures in the same units. It doesn't make any difference whether we convert both to mm Hg or to atm. We chose to convert to mm Hg. Volume is decreasing, so temperature must decrease; pressure and temperature are directly related (see Sec. 12.8 in ECOT), so since pressure is increasing, temperature must increase.

$$T_2 = T_1 \text{ (factor of volumes) (factor of pressures)}$$

$$T_2 = 310 \text{ K}\left(\frac{300}{400}\right)\left(\frac{912}{400}\right)$$

$$= 530 \text{ K}$$

$$= 530 - 273 = 257 \,^{\circ}\text{C}$$

Don't forget to reread the problem when you are finished and determine if you answered the question(s) asked. It is easy to forget to convert temperature in K to ˚C. Incidentally, you can see that the dominant factor in determining the new temperature in this problem is the change in pressures, not the change in volume.

If you need to convince yourself that this reasoning does work, solve the combined formula for T_2 and plug in the numbers.

Other Gas Laws

Two other gas laws are as closely related to each other as Boyle's and Charles' laws are in the combined law. Gay-Lussac's Law of Combining Volumes is stated in Sec. 12.9 in ECOT, and it's extension, Avogadro's Law is in Sec. 12.10. Let's work a simple problem illustrating Gay-Lussac's law.

Example 5: Ammonia, NH_3, is formed by the reaction of hydrogen, H_2, and nitrogen, N_2. If 6 L of hydrogen are allowed to react with enough nitrogen at constant temperature and pressure, how many liters of ammonia will form?

This is a stoichiometry problem — remember Chapter 9? Gay-Lussac's law says that when two gases react with each other (P and T constant) they combine in volumes that are simple whole number ratios. The molecules in chemical equations are in simple whole number ratios, aren't they? And it turns out that the ratio of volumes of gases is the same as the ratio of molecules. That's Avogadro's law. So if we write the equation, we can solve this problem.

Solution:

Naturally, since this is a stoichiometry problem, the first thing we need is a balanced chemical equation!

$$3 \, H_2 + N_2 \rightarrow 2 \, NH_3$$

According to Gay-Lussac, the equation says that 3 L of H_2 react with 1 L of N_2 to produce 2 L of NH_3. Using this simple whole number ratio of volumes, and knowing that we have enough nitrogen for the reaction, we can see that if 3 L of hydrogen produce 2 L of ammonia, then 6 L of hydrogen produce 4 L of ammonia.

Avogadro's Law goes one step further. Read the law in ECOT, and then look at the problem we just solved. We could have asked: How many moles of ammonia would be formed from 6 moles of hydrogen? The answer, of course, is 4 moles. This may seem obvious to us, but when Avogadro's Law was formulated in the early nineteenth century, it led the way to a better understanding of how atoms and molecules interact and ultimately to the concept of the mole.

Questions Answered

12.1   a. Liquids and solids; b. solids.

12.2   a. Propeller, football kicked end-over-end; spinning top;   b. power sander; guitar string after it's plucked; a gong after it has been struck; c. moving car, person walking, tennis ball after a serve (both translational and rotational).

12.3   Sodium chloride and potassium iodide: ionic bonding;
Water, ethyl alcohol, and propane: covalent bonding.
Since it takes more energy (higher temperature) to break bonds or otherwise separate particles which have attraction for each other, we must conclude that ionic attraction is much stronger than intermolecular attraction between the covalent compounds.

12.4   Half the sea level value, 760/2 or 380 mm Hg. (Be sure you have the correct units.)

12.5   They would have to be perfectly resilient, elastic spheres with no diameter or volume, losing no energy in their collisions, moving about continuously.

12.6   A Boyle's law problem:

$$V_1 = 100 \text{ mL}$$
$$V_2 = 95 \text{ mL}$$
$$P_1 = 760 \text{ mm Hg}$$
$$P_2 = ?$$

Volume is decreasing, so pressure must increase. To make that happen the volume factor must be greater than 1.

$$P_2 = 760 \text{ mm Hg}\left(\frac{100 \text{ mL}}{95 \text{ mL}}\right)$$

$$= 800 \text{ mm Hg}$$

12.7   A Charles' law problem:

$$V_1 = 10 \text{ L}$$
$$V_2 = 20 \text{ L}$$
$$T_1 = 25°\text{ C}$$
$$T_2 = ?$$

First convert T_1 to Kelvin: $T_2 = 25 + 273 = 298 \text{ K}$

Then: Volume is increasing, so temperature must be increasing. Multiply T_1 by a volume factor that is greater than 1.

$$T_2 = 298 \text{ K}\left(\frac{2 \text{ L}}{1 \text{ L}}\right)$$

$$= 596 \text{ K}$$

Converting to Celsius: $596 - 273 = 323 °\text{C}$

12.8   This is a pressure-temperature problem with volume constant. (See the spray can example in this section in ECOT.)

$$P_1 = 760 \text{ mm Hg}$$
$$P_2 = 300 \text{ mm Hg}$$
$$T_1 = 27 °\text{C} + 273 = 300 \text{ K}$$
$$T_2 = ?$$

To work this without using a formula, think about the spray can. What happens when the "empty" can is heated? The pressure of the gas remaining in the can increases, and the can explodes. That means that an increase in temperature brings about an increase in pressure — a direct relationship. So let's see: in this problem, pressure is decreasing by more than half, volume is constant, so temperature must also decrease.

$$T_2 = 300 \text{ K} \left(\frac{300 \text{ mm-Hg}}{760 \text{ mm-Hg}} \right)$$

$$= 118 \text{ K}$$

Convert to Celsius: 118 K − 273 = -155 °C.

12.9 The equation for the formation of water is

$$2 \text{ H}_2 + \text{O}_2 \rightarrow 2 \text{ H}_2\text{O}$$

From the equation we can see that the mole ratio of hydrogen to oxygen is 2:1, and we know from Gay-Lussac's law that the volume ratio of hydrogen to oxygen is 2:1 also.

12.10 The question says that $2\text{H} \rightarrow \text{H}_2$. The mole ratio is 2:1. Thus the volume ratio must be 2:1 also, so 10 L of H atoms would become 5 L of H_2 molecules, assuming temperature is constant.

12.11 Nitrogen. Since we know that air is somewhat soluble in water, then the higher partial pressure of nitrogen in the atmosphere, according to Henry's Law, would produce a higher concentration of nitrogen than oxygen dissolved in the water.

12.12 Pressure; surface area of contact between the N_2 and the water.

12.13 Water vapor is the fourth gas. You can confirm this by placing a drinking glass in the freezer for a few minutes and then breathing on it. Water vapor, along with carbon dioxide, is a product of the metabolism of glucose.

12.14 Water would rise in the bottle as the cold towel cools the air in the bottle. As the gas cools, its molecules will move more slowly and will occupy less space inside the bottle (contract). Atmospheric pressure outside the bottle pushes water into the bottle. Try it!

Supplementary Exercises

S1. A gas is collected in a 2-L cylinder at 720 mm Hg. A piston is used to increase the pressure on the gas to 1440 mm Hg. What is the new volume? (Temperature is constant.)

S2. A gas has an initial volume of 25 L at atmospheric pressure and 25 °C. The pressure is lowered to 350 mm Hg and the gas is heated to 100 °C. What is the new volume?

S3. A gas is collected at atmospheric pressure and room temperature, 25°C. The measured volume is 5 L. At what pressure will the volume occupy 10 L and the temperature measure -10 °C?

S4. A weather balloon having a volume of 250 liters at sea level (P = 1 atm; t = 25°C) is released carrying instruments and a transmitter to radio data back to base. When the balloon has risen to programmed height (14 km) where the pressure is 75 mm Hg and the temperature is -53°C, to what volume will it have expanded?

S5. The pressure gauge on a welder's oxygen tank registers "0 lb/in^2," indicating to the welder that the tank is empty. Is the tank empty? Explain. (Hint: Recall Exercise 7 in ECOT.)

S6. Assuming temperature and pressure are constant, how many liters of hydrogen chloride gas can be produced from 12 L of hydrogen and 6 L of chlorine? Is HCl the only gas in the container after the reaction?

Where You Might Goof Up...

- Forgetting to change local temperature into absolute temperature.

- Being unable to use your own body, while in the standing position, to demonstrate translation, rotation, and vibration.

- Incorrectly listing translation, vibration, and rotation in order of increasing need of energy for activation of the mode of vibration.

- Being unable to state Boyle's, Charles', Avogadro's, Dalton's, and Henry's Laws in words.

- Being unable to use Boyle's, Charles', Avogadro's, and Dalton's Laws to solve numerical problems.

- Improperly listing the major gases found in our atmosphere.

- Being unable to describe the process of nucleation and how it relates to Henry's Law.

- Inadequately outlining the process of respiration because you omitted either diagrams or equations.

CHAPTER 13

Surfactants

SOAPS AND DETERGENTS: CLEANING UP WITH CHEMISTRY

Continued from Chapter 10. . .

"At least," Angela reconsidered, "I know how to show that it could be magnesium oxide or calcium oxide. I need the bar of soap from the rest room, a vial or a small bottle with a lid, a knife, and some water."

"There's some detergent under the sink," Jamal offered.

"No," Angela said. "It's gotta be soap. Detergent won't work."

They gathered the materials together, and Angela scraped some soap into a pill vial. She added a few milliliters of water, capped the vial, and shook the container.

"There," she said, holding the vial up for inspection.

"Angela's good," said Jimmy with mock seriousness. "She's made soap suds."

Angela laughed and scraped a small amount of the mysterious dust onto a piece of paper. "Let's see what happens when I add the dust to the soapy water."

She added some dust and shook the vial. Suds disappeared, and clumps of precipitate appeared.

"Voilà! Soap scum forms when the calcium or magnesium ion replaces the sodium ion of the soap and forms a precipitate." she exclaimed. "We've shown that the dust may be magnesium oxide or calcium oxide."

"Very good," Jimmy said. "Both tests you two have done have indicated that this white powder may be CaO from the lime kiln. But have you proven that it is CaO?"

Jamal and Angela looked at each other.

"No," Jamal said. "We've only pointed the way for further research to go."

"Right," Angela agreed. "But we can be fairly certain that we could ask the lab at the kiln to confirm it for us. They test their product all the time."

"Exactly," Jimmy said. "I'll take a sample over to the kiln and get them to run it through one of their instruments, just to be on the safe side. Then we can clean up this mess."

Chapter Overview

• A demonstration is described with which you can confuse a friend. Practice makes perfect.

• Why bugs don't sink is related, and density is discussed.

• Surface tension, caused by the difference between the outside and the inside of a body of water, is described.

• The two-facedness of soaps, detergents and surfactants is pointed out and how they clean is explained.

• Micelles, colloids and how to detect them is recounted.

• The way soap is made is described.

• The causes and some solutions to the problem of hard water are given.

- A new class of organic compounds, esters, is introduced with methods of synthesizing them.

- How to avoid the ring in the bathtub is suggested.

For Emphasis

Density

Table 13.1 of ECOT lists the densities of many different kinds of materials: elements, compounds and mixtures. If the substance is pure (an element or compound), the density is an important invariant property. Like temperature and voltage (see Chapter 11 in this *Guide*), density is an *intensive* property. The density of aluminum is 2.7 g/cm³, whether we have 1 kg or 1 ton of Al. However, if the substance is a mixture such as gasoline, the density can vary from batch to batch. Densities are easily measured if the substance is a liquid or a cubic or rectangular solid. In the case of a liquid, it is necessary only to measure the volume of a known weight of liquid and perform a calculation such as the one in the example in Section 13.1 of ECOT. If a solid has a regular shape such as a cube, its sides can be measured, its volume then calculated and the solid weighed; knowing the mass and the volume allows the calculation to be performed. Even if it is round or dowel-shaped, diameter and length can be measured and the volume determined. What if the solid is irregular?

Colloids — The Tyndall Effect

The Tyndall effect, described in Section 13.5 of ECOT, is all around us. We can see it in the dark as well as with the assistance of clouds in the daytime. We see the Tyndall effect while sitting in a darkened room with the shades drawn. If there is any light outside the window, and the shades aren't tight, a white path defined by the interaction of the leaking light with dust particles appears. This Tyndall effect is observed at right angles to the beam of light; we must be looking across the beam. All of us have seen Tyndall scattering every time we have gone to the movies. The next time you go to the movies, look up and see the projection beam reach from the projection room to the screen, its light scattered by the dust in the theater.

Esters and How to Draw Them

An ester is a product of the reaction of an alcohol and a carboxylic acid. Water is the other product. Figure 13.13 in ECOT shows an acid catalyzed esterification, which, simply put, is the production of an ester (ethyl acetate in this case) in the presence of an acid catalyst. A mineral acid like hydrochloric acid or sulfuric acid is often used as a catalyst. Notice the name of the ester. The first name of an ester comes from the alcohol from which it was formed. "Ethyl" is from ethanol or ethyl alcohol. The second name is the name of the carboxylate ion of the carboxylic acid. Acetic acid is the carboxylic acid; acetate is the name of the carboxylate ion. Note in the following equations in ECOT that although the first name of the ester is the name of the alcohol, by convention the acid portion of the molecule is drawn first. The general formula of an ester is RCOOR', and the general reaction is

carboxylic acid   alcohol   ester

To draw some other esters, let's first make a list of some simple alcohols and some simple carboxylic acids. Fill in the spaces in the table on the next page.

Alcohol	Formula	Acid	Formula
Methyl alcohol		Formic acid	
Ethyl alcohol		Acetic acid	
Propyl alcohol	$CH_3CH_2CH_2OH$	Propionic acid	

Using these compounds as starting materials, draw formulas of esters using the general structural formula above as a pattern. For example, starting with methyl alcohol, you could draw methyl formate, methyl acetate, and methyl propionate.

Look up the formula of benzoic acid in Chapter 13 of ECOT, and draw some more esters.

Distinguishing Detergents

All three surfactants discussed in ECOT have "double-ended" molecules: one end that is hospitable to water and the other attracted to the nonpolar grease and soil. This is an ideal situation in which to construct an information table such as the one in Chapter 4 of this *Guide*; the vertical column at the left of the table should read: *Anionic, Cationic, Nonionic*. The horizontal labels across the top might read: *Structural, Other Examples, Applications, Comments . . . etc.*

A practical, large-scale example of the effect of detergents on surface tension is easy to do, using capillary action. Capillary action is not the only mechanism for this phenomenon, but it is one of the important ones. Obtain a drinking straw; a clear one is better than an opaque one, a thin one is better than a fat one. Dip the straw in a glass nearly filled with water and slowly withdraw the straw part way. Notice that the water inside the straw stands a little higher (a few millimeters, maybe) than the water outside the straw. Now try the same experiment with a detergent solution. Does the level of the liquid in the straw stand higher in the water in the first experiment or in the detergent in the second?

Questions Answered

13.1   The density of aluminum is 2.7 g/cm^3; the density of lead is 11.4 g/cm^3. Use the volume and the density of each metal to determine their respective masses:

$$\text{density} = \left(\frac{\text{mass}}{\text{volume}}\right), \text{ so mass} = \text{density x volume}$$

for aluminum, mass = 2.7 g/cm^3 x 10 cm^3 = 27 g
for lead, mass = 11.4 g/cm^3 x 2 cm^3 = 22.8 g

The sample of aluminum is heavier. Since density is an intensive property, it doesn't matter how much of each metal we have. Lead is denser.

13.2   Heat the water. The surface tension of water decreases on heating from room temperature to the boiling point. As the temperature is raised, thermal agitation increases which decreases the effect of the attractive forces under the water's surface. Or, you could add detergent to get the same effect at room temperature.

13.3   A long hydrocarbon chain is the hydrophobic portion; the ionic end is hydrophilic (the structure drawn in Section 13.3 of ECOT is the carboxylate ion).

13.4 Surfactants, detergents, and soaps are all surface-active agents that lower surface tension. Detergents are good cleaning agents that may vary in structure, but all have one molecular structure in common: a hydrophobic carbon chain that resembles a molecular "tail" and a hydrophilic "head". Since oil spreads out over a surface and doesn't diffuse into the water, it could be considered a surfactant.

13.5 Shine a beam of light from a flashlight through the mixture. If you can't see the beam, the mixture is a solution; if you can see it, the mixture is a colloid.

13.6 Detergents clean by converting dirt into micelles, which are suspended in the water and can be rinsed away.

13.7 The result would be methyl propionate.

```
      H   H      O
      |   |     //
  H―C―C―C       H
      |   |    \  |
      H   H    O―C―H
                  |
                  H
```

13.8 All triglycerides are triesters of glycerol. They differ by the identities of the various fatty acids that combine with the glycerol.

13.9 If you shake the soap in hard water first, curds form immediately when soap reacts with the calcium ion, Ca^{2+}, in tap water. Addition of distilled water at this point would make no difference.

13.10 Hard water ions, such as Ca^{2+} and Fe^{3+}, are rinsed off the resin and go down the drain.

13.11 Since detergents remain dispersed even in the presence of hard water ions such as Ca^{2+}, the detergent would not form curds in tap water. The demo would show the same result for distilled water as for tap water.

13.12 Phosphates were originally added to detergents as builders (water-softening agents). They exhibited low cost, low toxicity, and an ability to bind with hard water ions. But the downside is that they were superb algae nutrients. When algae grow too fast (algal bloom), they prevent atmospheric oxygen from mixing into bodies of water, and when the algae die, the process of decaying removes even more oxygen from the water. The loss of dissolved oxygen in the water led to the deaths of many aquatic animals, so phosphates were removed from detergents.

Supplementary Exercises

S1. An astronaut returned from Mars with a small piece of a strange, beautiful mineral which was irregularly shaped. Given a glass, water, a saucer, and a balance, how would you determine its density?

S2.   A student is given a set of wooden blocks 2 cm on a side to weigh in a physical science lab. The data collected are as follows. a. ash, 5.6 g; b. ebony, 9.76 g; c. African teak, 7.8 g; d. lignum vitae, 10.5 g; e. dogwood, 6.1 g. Which will not float?

S3.   Name the esters below.

a.

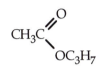

b.

$$CH_3C \overset{O}{\underset{OC_3H_7}{\diagdown}}$$

$$C_{17}H_{35}C \overset{O}{\underset{OC_3H_7}{\diagdown}}$$

S4.   Name the esters that would be synthesized from the following pairs of reactants:

a.   benzoic acid + methyl alcohol

b.   butyric acid + propyl alcohol

c.   octyl alcohol + propionic acid

d.   propionic acid + butyl alcohol

e.   oleic acid + capryl alcohol

S5.   From the point of view of a Ca^{2+} ion, what is the *chemical* difference between a soap and a detergent?

Where You Might Goof Up...

- Inverting the equation for the definition of density.

- Not being able to account for the phenomenon of surface tension using words and diagrams.

- Describing the effect of a rise in temperature on surface tension inaccurately.

- Confusing the words *hydrophobic* and *hydrophilic*.

- Not recognizing the Tyndall effect the next time you are in a movie theater.

- Allowing the general formula, $R-\overset{\displaystyle O}{\underset{}{C}}-OR'$ to go unrecognized as a generic ester.

- Distinguishing inadequately among soaps, detergents and surfactants.

- Distinguishing inadequately among cationic, anionic and nonionic detergents.

- Being unable to describe at the molecular level how the two different ends of a detergent clean your laundry.

CHAPTER 14

Chemicals, Pollution and the Environment

THE MEANING OF POLLUTION

The aldehyde plant washed its spent mercury catalyst into Minamata Bay where it would be carried harmlessly out to sea. Plankton took in the Hg in low concentration, small fish ingested the plankton concentrating the Hg, larger fish ate the smaller fish concentrating the Hg still further, then the larger fish were harvested by the fishermen who, after removing entrails and throwing them in the garbage where neighborhood cats foraged, consumed the fish. One afternoon, Fumiyo came home from her grade school to find her cat walking awkwardly in the yard. She called out but the cat, instead of coming to her, danced around in tight circles and fell over, panting, glassy-eyed; it died before Fumiyo could get it into the house. After a time, Fumiyo's aunt bore a child having only three fingers who never learned to walk, and a little later Fumiyo's friend got a new brother who lived only a few weeks, which was considered a blessing. Fumiyo is grown up, but will never marry. After all, who wants to have children with a with a person exposed to "The Disease of the Dancing Cats"?[*]

Chapter Overview

- A demonstration is described which encompasses some simple combustion chemistry, some acid-base chemistry with indicators *(deja vu)*, the scientific method and an explanation for acid rain.

- The steps in the chemical formation of acid rain are given and discussed.

- The structural parts of the earth are listed and diagrammed.

- An operating definition of pollution is given.

- The contribution of fossil fuels to air pollution is described.

- The major pollutants are listed.

- A distinction between *primary* and *secondary* pollutants is made and examples given.

- Photochemical smog is treated, necessary reactants listed, and chemical processes detailed.

- The chemical reactions of acid rain are detailed.

- Options for abatement of air pollution are listed.

- The four forms of water pollution and their sources are discussed.

---

[*] Smith, W.E. & Smith, A.M. *Minamata.* New York: Holt, Reinhart & Winston (Aiskog-Sensorium) 1975. (This is a detailed story of the "Disease of Dancing Cats" in words and pictures. This publication has given the city and the factory a permanent place in environmental history.)

- Eutrophication as an overabundance of nourishment is discussed.

- Pesticides are examined and the desirability of *biodegradability* is pointed out.

- The problem of the *disposal of waste,* from the exotic to the ordinary, is discussed.

For Emphasis

Reviewing the Arithmetic of Chemistry

Chapter 6 of ECOT introduced you to some of the calculations you must make to discuss pollution *quantitatively.* You should review concentrations such as molarity, parts per million, w/w %, and go back over dilution calculations.

In Chapter 9 you did acid-base calculations. Among those were pH and its relation to concentration of the H^+ ion. Review that chapter.

Learning Chemical Equations

This chapter is rife with chemical equations which describe the reactions of pollution. Chemical equations should be addressed like any information you want to learn. Like the habit that you undoubtedly have now of "quizzing the book," make a habit of quizzing every reaction you encounter. For example, take the first equation in Chapter 14 in ECOT in the Demonstration. It is the reaction of sulfur dioxide with water.

$$SO_2 + H_2O \rightarrow H_2SO_3$$

This is an easy one, but don't let it slip by you. Never simply read an equation. *Always* write the equation down, in symbols first, and then in words. Ask the equation all the questions you can think of. Then find the answers, if possible.

- Where did the sulfur dioxide come from? (Paragraph above the equation in ECOT — burned a match.)

- So how does H_2SO_4 form? (In the next paragraph. It forms when SO_3 dissolves in water.)

- What's that equation? (Follows that paragraph — $SO_3 + H_2O \rightarrow H_2SO_4$)

- But where does the SO_3 come from? (Paragraph below the equation we started with. ECOT says "SO_2 reacts with the oxygen of the atmosphere and slowly oxidizes to sulfur trioxide.")

- What's the equation for that oxidation? (It's not in this Section of ECOT, but you can figure it out. $2SO_2 + O_2 \rightarrow 2SO_3$. Looking ahead, it's found in Section 14.4.)

- What's the outcome of all this? (Both acids dissolve in rain water — acid rain is an outcome.)

- What's the pH of acid rain? (ECOT says, "We'll have more to say about acid rain and other forms of pollution later in this chapter." Later, then.)

There are more questions you could ask. When you feel you have gleaned a considerable amount of information from the exercise, write all the equations you have just explored one more time. Later in the chapter in Section 14.4, for example, you will encounter some of the same equations again, equations which you now know a great deal about, and you have some new ones to quiz.

In Section 14.6 sulfur dioxide comes up again, this time in the removal of SO_2 by a scrubber.

$$SO_2 + Mg(OH)_2 \rightarrow MgSO_3 + H_2O$$

and

$$MgSO_3 \rightarrow SO_2 + MgO$$

Now you have more equations to quiz. "Can the removed SO_2 be used to produce sulfurous acid commercially?" "Could $Ca(OH)_2$ be used as well?" "Wouldn't we have a reaction similar to the reaction of SO_2 and $Mg(OH)_2$ if H_2SO_3 were to react with $Mg(OH)_2$?" "Wouldn't that be an acid-base reaction?" "Does that mean . . ."

You get the idea. You will ask some questions that don't have easy answers, or for which you can't find an answer in ECOT. Remember in the introduction to this *Guide* it says that you should be alert, sitting in the front of the class. Ask your instructor good, considered questions, and you will be noticed, even in a large lecture section. And the best part is that you'll know the material.

You will find lots of equations in the chapters to come. Some of them look more complex and have structural formulas given. Apply the same techniques, and you'll learn them.

Questions Answered

14.1 a. CO_2; b. H_2O

14.2 This is a "Think, Speculate, Reflect, and Ponder" question. (Refer to Section 9.12 in ECOT.)

14.3 +4. (In FeS_2, each sulfur has a –2 charge. Therefore, the valence of Fe must be +4.

14.4 a. $N_2 + 2\,O_2 \rightarrow 2\,NO_2$;
b. Analyze the question: We want a single equation in which N_2 reacts with O_2 and one of the products is O_3. There is no equation with these properties in Sec. 14.4. However, we can construct one by adding together equations that *are* in that section and using the equation in a. of this section.

Let's gather the equations. In a above, we find the reactants required:

$$N_2 + 2\,O_2 \rightarrow 2\,NO_2$$

Since the product we want is O_3, find equations on p. 339 of ECOT that will give us O_3. Add them together algebraically:

$$\begin{aligned}
NO_2 &\rightarrow NO + O \\
\underline{O + O_2} &\underline{\rightarrow O_3} \\
NO_2 + O_2 &\rightarrow NO + O_3
\end{aligned}$$

Notice that we have subtracted O from both sides. Now add the equation in a to the one we just obtained, multiplying it by 2 to be able to subtract 2 NO_2 from both sides:

$$\begin{aligned}
N_2 + 2\,O_2 &\rightarrow 2\,NO_2 \\
\underline{2NO_2 + 2\,O_2} &\underline{\rightarrow 2NO + 2O_3} \\
N_2 + 4\,O_2 &\rightarrow 2\,NO + 2\,O_3
\end{aligned}$$

14.5 Volcanic eruptions.

14.6 Scrubbing.

14.7 Summer. Since intense sunlight contributes to ozone formation, the longer days of summer would produce higher concentrations of ozone.

14.8 Harms: cataracts, skin cancer; Benefits: formation of vitamin D.

14.9 They are all halogenated hydrocarbons. They all absorb UV rays, decomposing to a halogen atom and other products that catalyze the breakdown of ozone (O_3) into oxygen gas (O_2), thus destroying the ozone layer.

14.10 Thermal: Gases are less soluble in warm water than in cold; Sedimentary: Particles of insoluble organic chemicals can cause a drop in the level of photosynthesis in plants in the water. They also carry chemical and biological pollutants with them.

14.11 BOD: high; DO: low; DOD: high.

14.12 An aquifer is an underground deposit of water, usually in porous rock formations. Aquifers participate in dynamic water supply systems: in drought, water flows from them into lakes, streams and rivers; in flood times they are replenished. The water is drawn for use from wells.

14.13 In scrubbing, water is used to remove harmful chemicals; in air stripping, air is used to remove volatile solutes from water.

14.14 Parathion, malathion, and diazanon are used because they do not last indefinitely in the environment the way DDT does. However, DDT is less toxic than they are in the short run.

14.15 This would not be better. Either way, the thinner gets into the environment, polluting it. A better option would be to take the paint thinner to a collection center for disposal.

14.16 Use what you have learned in this chapter to state your opinion and make a well-reasoned argument to support it.

Supplementary Exercises

S1. Convert the following concentrations. Review Chapter 9 in ECOT and in this *Guide* as necessary.

  a. 20 ppm to mg/L:

  b. 20 ppm to % by weight:

  c. 400 mg/L of As to molarity of As:

  d. 0.025 M Ca^{2+} to ppm:

  e. 2 ppm to ppb

S2. In Table 14.4, which source of air pollution has the highest percent VOCs?

S3. Without looking back at Chapter 14 in ECOT until you have tried this question, write equations for the following:

a. Solution of sulfur dioxide in calcium hydroxide.

b. Solution of carbon dioxide in water.

c. Reaction of sulfur trioxide with water.

d. Conversion of ozone to oxygen.

e. Production of sulfur dioxide from iron pyrite.

S4. You are driving down the road eating a candy bar, and you are tempted to throw the wrapper out the window, thinking that "just one" wrapper won't make any difference. If just 1% of people with cars discard 1 candy wrapper each week for 1 year, how many candy wrappers will be discarded on public highways in 10 years? Go back to Section 7.1 of ECOT for data.

S5. Which of the following are primary pollutants, secondary pollutants or neither?
a. NO;   b. O_2;   c. H_2SO_4;   d. S;   e. O_3.

S6. A chemist determines that the $[H^+]$ of rain falling in an area is 1×10^{-4} M. Is this acid rain? Explain.

Where You Might Goof Up...

- Defining pollution too narrowly.

- Failing to discuss pollution abatement as a series of trade-offs.

- Being unable to list the various sources of water pollution.

- Describing inadequately the role of the sun in photochemical smog.

- Being unable to give the equations for the reactions responsible for acid rain.

- Neglecting to consider eutrophication as an *overabundance* of nourishment.

- Being unable to describe how a secondary pollutant is generated.

- Forgetting that the long-term use of a pesticide leads to its becoming ineffectual.

- Overlooking the fact that *waste disposal* is becoming the largest pollution problem, especially in industrialized society.

CHAPTER 15

Fats and Oils

THE TRIGLYCERICIDES WE EAT

It's so important that its design fills several pages of the Federal Register, its specifications agreed upon after years of research, every line, every space determined by a committee of experts. To see it is to appreciate the spare beauty of the graphic design and the standardized access to information that it provides us.

What is it? No, it's not a new spacecraft; it's not a new wireless phone to provide everyone access to the Internet. It's just a nutrition label, but its use, required by the Nutrition Labeling and Education Act, revolutionized the way we receive nutrition information about the foods we eat. Before the law went into effect in 1993, food manufacturers could decide for themselves what constituted a serving size. In high fat potato chips, for example, in order to avoid the perception of "unhealthy," a manufacturer might define the serving size as five chips. The new law meant not only that we get the same kind of information from every label on every package of food, but serving sizes are standardized, making them more realistic, as are nutrient claims, such as "fat free," "light," and "low calorie," for example.

Nutrition and public policy experts and product assessment and policy enforcement mavens at the Food and Drug Administration and the U.S. Department of Agriculture Food and Safety Inspection Service worked together to arrive at a regulation that was reasonable, enforceable, that met the needs of consumers as well as those of the food industry. They sought input from health professionals, consumer groups, industry representatives, nutrition experts, scientific studies. The result was a good compromise that generally had the support of both consumer groups and the food industry.

One doesn't have to have access to the Federal Register to see the label. By law (the one printed in the Federal Register on January 6, 1993), it's on every food package on the market.

Chapter Overview

- An experiment using common oils and drugstore iodine is detailed for you to try.

- Two different kinds of triglycerides, fats and oils, are distinguished.

- The chemical difference between fats and oils is shown.

- The chemical and physical effects of the presence of double bonds in some triglycerides are considered.

- The implication of polyunsaturation in dietary regimen, spoilage and paint drying is put forward.

- Catalytic hydrogenation is covered.

- Cholesterol is examined.

- Reactions of fats with oxygen are uncovered, and antioxidants are identified.

- *Cis-trans* isomerism and why oils melt at a lower temperature than fats are explained.

- Healthy triglycerides and unhealthy triglycerides are examined.

- The popular omega fatty acids are detailed.

• Adipose tissue, the stuff we like to minimize, is defined.

For Emphasis

Fats, Oils, and the Double Bond

In this chapter, much is made of the fact that the biggest difference between fats and oils is that oils have double bonds and fats do not. Let's look further into the nature of the double bond. If you refer to Figure 6.15 (p. 134 of ECOT), you will see a diagram of ethylene, the simplest hydrocarbon having a double bond. Using your gumdrop kit, construct ethylene. For construction, two points should be emphasized:

• All six atoms (4 H's and 2 C's) are in the same plane; (this is a "two-dimensional" molecule). After you finish the assembly, lay the model on a flat surface; all six atoms should touch the surface.

• There must be two toothpicks connecting the carbons together. This correctly forbids rotation around the double bond.

In ethane (single bond between the carbons) there *is* rotation around the bond; the methyl groups of the model can spin like propellers on a model airplane. The higher the temperature, the faster they spin. In that way, the gumdrop model is a *good* model in that it predicts the truth for ethane. With double bonds in position on the model as is the case with ethylene, the substituents on each carbon cannot rotate; the gumdrop model is a good one for ethylene as well. Now assemble two gumdrop methyl groups — CH_3. Substitute each of them for a hydrogen atom on each end of the ethylene (not on the same carbon). On a test, you *might* call this molecule 2-butene. If each of you in a study group has done this, your molecules probably do not all look the same although you followed the instructions correctly. Some of the models will have both of the methyl groups above the double bond, some of the models will have one methyl group on each side, and some will have both below the double bond. All of these models are 2-butene, but their physical and chemical properties are different, not very different, but different. Thus we must distinguish them further. Let's look at the structures below:

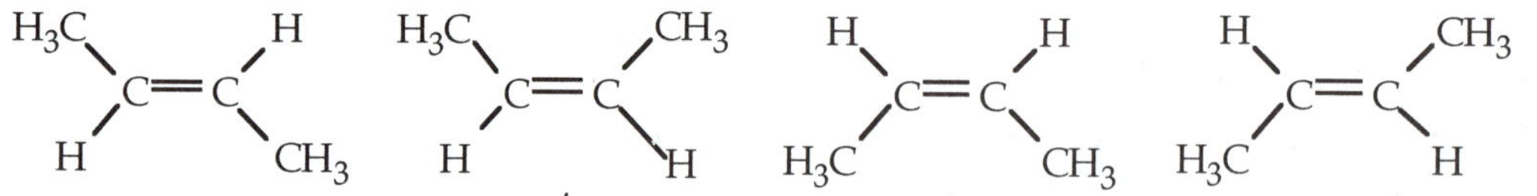

There are really only two butenes here. You can prove this to yourself by flipping the models over. In other words, the *trans* at the beginning of the line is the same compound as the *trans* at the end. The reason we can have such a variation in geometry is because of the double bond. Visualize a substituted ethylene with an imaginary line drawn across the molecule through the double bond:

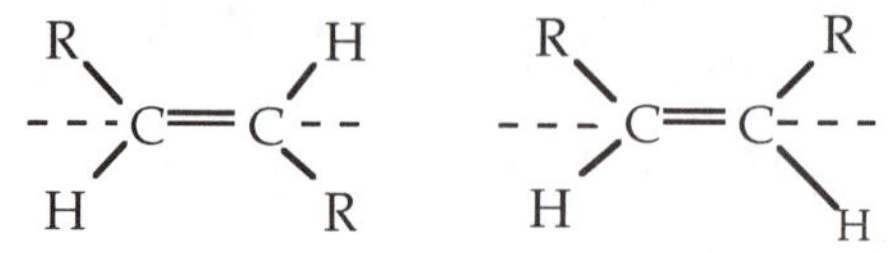

If the substituents are on opposite sides of the line coincident with the double bond, the structure is *trans*; if they are on the same side, the structure is *cis*. Be assured that none of the H's in the *methyl groups* of the models you made need touch the plane; the C's of the methyl groups *are* in the plane however. The conclusion to all of this is that one must name the compound whose model you made either *cis*-2-butene or *trans*-2-butene. If you omit this, the name is ambiguous and usually considered incorrect because it is incomplete.

The naturally occurring fatty acids are all *cis*. For oleic acid, the double bond comes after the ninth carbon (counting from the carboxyl end). You and your study group could share gumdrops and make the model of oleic acid:

oleic acid

Once you make a *cis* double bonded compound larger than a butene, you can see how difficult is would be for the molecules to pack in an orderly fashion in the solid state. They just do not touch each other enough; thus we have to cool them off more for them to become solid. Table 7.2 of ECOT shows how the boiling point of alkanes drops with increasing branching; refer to Chapter 7 of this study guide for a detailed discussion of why it changes this way. The same thinking applies to the comparison of saturated and unsaturated fatty acids. The saturated fatty acids have long straight chains much like the straight-chain alkanes; the unsaturated fatty acids are bent (*cis*), so they cannot line up with each other for any great length and mutually attract. The forces acting here are less than if the molecules were lying close all along each other's lengths.

Questions Answered

15.1   Oils to fats.

15.2   Most highly saturated has the lowest iodine number: coconut oil; most highly *un*saturated has the highest iodine number: linseed oil (fish oils run a close second).

15.3   Look at the fatty acids in Table 15.1. The fatty acid side chains in glyceryl trilinoleate are linolenic acid, which is polyunsaturated. Therefore, it would be a good candidate for candy bar production. The same is true for glyceryl trioleate (identify its fatty acid in the table). The least useful is glyceryl tripalmitate because it contains no double bonds and would tend to give the chocolate a waxy, brittle consistency.

15.4   Red meat and animal products contain significant quantities of cholesterol and tend to have large quantities of saturated fats that the liver converts into even more cholesterol.

15.5   Don't add the antioxidant, because a drying oil works by reacting with oxygen in the air to form a hard surface. Although an antioxidant might protect the oil while it's in the can, its continued presence in the applied oil would prevent the required oxidation from occurring.

15.6   a. incapable (2 identical groups on the same carbon); b. *cis*; c. *trans*; d. incapable (2 identical groups on the same carbon).

15.7   A polyunsaturated triglyceride contains double bonds that cause the fatty acid chains to kink which interferes with efficient packing and reduces the attractive forces between the triglyceride molecules. Weaker attractive forces can be overcome by less energy, so the polyunsaturated triglyceride will melt at a lower temperature. By hydrogenating the polyunsaturated triglyceride, the double bonds are removed and the packing of the fatty acid chains becomes more efficient. More efficient packing leads to stronger attractive forces that require more energy to overcome, so the saturated triglyceride will melt at a higher temperature.

15.8 Omega-3.

15.9 15 pounds $\left(\dfrac{3500\ \text{Cal}}{1\ \text{pound}}\right) = 52{,}500\ \text{Cal}$

Supplementary Exercises

S1. When oleic acid is hydrogenated, stearic acid results. Using Table 15.4 for formulas, write a balanced equation for the hydrogenation of oleic acid.

S2. Calculate the number of grams of hydrogen needed to hydrogenate 100 g of oleic acid.

S3. Write the balanced chemical equation for the reaction between iodine and oleic acid.

S4. Calculate the number of grams of iodine needed to react with 100 g of oleic acid. *The Merck Index*, item #6788, gives 89.9 as the experimental iodine number for oleic acid. How does this compare with your theoretical calculations?

S5. Using Table 15.2, which animal fat and vegetable oil have the highest percentage of monounsaturated fatty acids? The lowest percentage of saturated fatty acids? List the actual percentages for each.

Where You Might Goof Up...

- Incorrectly sketching the general structural diagram for triglycerides.

- Improperly writing a balanced chemical equation for the catalytic hydrogenation of an alkene or for the reaction of elemental iodine with an alkene.

- Failing to realize that the double bond in any molecule is a reactive center.

- Failing to distinguish between a saturated and unsaturated fat by looking at the structural formula.

- Being unable to identify the structural formula of cholesterol among those of other compounds of biological interest such as, fatty acids, triglycerides, alkyl alcohols and inorganic salts.

- Making mistakes while discussing the chemical and health implications of a high *vs.* a low iodine number as applied to triglycerides.

- Neglecting the implications for cholesterol formation from saturated fatty acid ingestion as well as from ingestion of cholesterol directly from animal fat.

- Being unable to construct models of *cis* and *trans* forms of compounds containing double bonds.

- Not understanding the role of oxygen in drying oils.

- Failing to predict, upon inspection of structural diagrams, whether geometric isomers are possible for certain compounds.

- Inadequately explaining at the molecular level why unsaturated fats are liquids at room temperature and saturated fats are solids.

- Failing to identify the omega-3 carbon on a fatty acid.

- Being unable to discuss the survival value of energy storage in our bodies.

CHAPTER 16

Carbohydrates

FOOD FOR THOUGHT

At the vineyards of Amun west of the Nile, Pharaoh's vintner, Seph-na-tet, took one last look at the vines now empty of their deep blue grape crop. The harvest had gone well. Soon the feast of Geb, god of the Earth, would begin, and the royal family would open some jars of last year's wine, now aged and (Seph-na-tet hoped) perfect. The common people celebrated the harvest with beer made from wheat and barley. It was a festival for the entire nome, and everyone was preparing for the celebration.

Seph-na-tet thought about last year's harvest and the wine he was preparing to deliver to the royal family. He often marked the birth of a new wine by events in his own family. Last year, at first fermentation time, his daughter had been born. He remembered watching the vats overflowing with the foam of new wine, as the news of her birth was brought to him. A second fermentation was quieter, sealed in wine jars with a small hole in the seal so the jar wouldn't explode. (Glucose, in the presence of microbes such as yeast, is oxidized to ethanol. Carbon dioxide is given off, and the pressure builds up. $C_6H_{12}O_6 \rightarrow 2C_2H_5OH + 2CO_2$)

During the year there was almost a disaster as mud seals on a few of the jars of Seph-na-tet's precious product were broken by a careless worker. He knew the wine in those jugs would be worthless —just vinegar now. (Further oxidation of ethanol to acetic acid: $2C_2H_5O + O_2 \rightarrow 2CH_3COOH + 2H_2O$). But the rest of the seals were intact. He looked them over one last time before the workers loaded the jars on oxcarts for the long trip to the palace. Each seal had the name of the royal family and told the story of the wine. In his last tasting of the wine Seph-na-tet had pronounced the wine "very good," and the seal proclaimed, "Very good wine of the house of Amun from the work of the vintner, Seph-na-tet." It was truly wine for a king.

To be continued in Chapter 22 . . .

Chapter Overview

• A demonstration is suggested by which you impress your friends because you can tell the difference between two common foods which look the same without eating either one.

• Our fuel supply, glucose, is identified.

• Mono-, di-, and poly-saccharides are defined.

• Two groups from Chapter 6, aldehydes and ketones, are revisited and shown to be important in biochemistry and nutrition.

• The right- and left-handedness of molecules (chirality) is explained and a technique for detecting this is described.

• Polarized light is explained, and how it behaves toward glucose, fructose and their mixture in solution is described.

• The cyclic structure of glucose is elucidated.

• The difference between starch and cellulose is explained and its dietary implications are enlarged upon.

- Enzymes and carbohydrate digestion are discussed.

- Lactose intolerance, occurring mostly in adults, is described.

- The idea of enzymatic "fit" is developed.

For Emphasis

Aldehydes and Ketones

A simple way to think about the difference between an aldehyde and a ketone is to visualize the structures below. Whether a compound is an aldehyde or a ketone depends on the value of X:

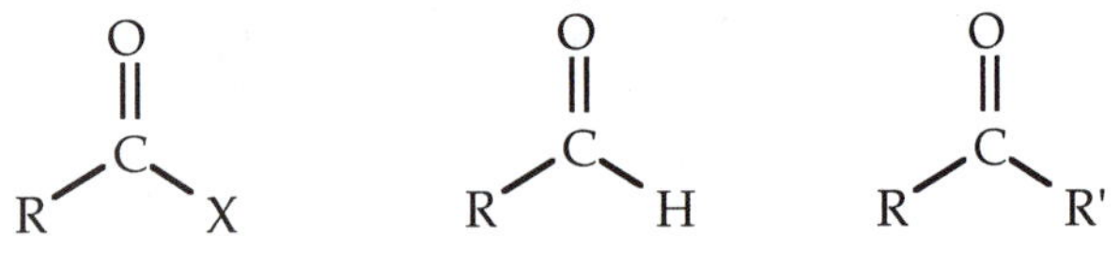

aldehyde   ketone

R' can be any other organic fragment such as —CH_3, —CH_2CH_3, a benzene ring, or even R itself. Although aldehydes and ketones can be isomers of each other such as acetone (CH_3COCH_3) and acetaldehyde (CH_3CH_2CHO), they can be very different chemically.

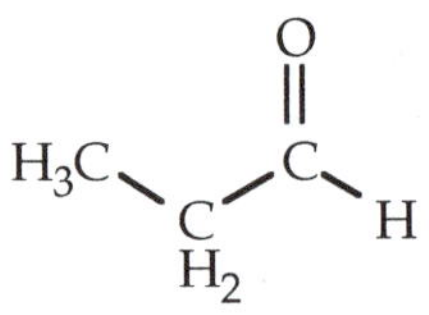

acetaldehyde 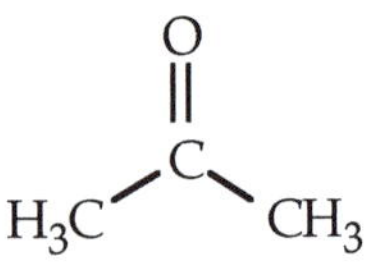   acetone

Acetaldehyde is very easily oxidized (so easily that it cannot be kept in a stockroom in a tight bottle without being oxidized in a short while), while acetone is very stable, being used for paint thinner, nail polish remover and glues among other things.

Enantiomers

The gumdrop model kit will help significantly in the understanding of the stereoisomerism of both a simple tetrahedral structure and the sugars. Turn to Figure 16.2 in ECOT. Select a black gumdrop for the central carbon atom shown in the figure; select four other gum drops, *each one of a different color*. Assemble this molecule with tooth picks. Now build another starting with a black central carbon and the same four colors you used on the first model. You have made either two identical molecules or two enantiomers. It is easy to tell which: Lay the models on the table next to each other with the same three colors of both models touching the surface you are working on; the fourth color (same on both models) will be extended into the air above the carbon. Rotate the models (without removing them from the flat surface) until the same color on both models is *away* from you. Supposing that the gumdrops on the table are red, yellow and green, and an orange one is above the carbon in the air — you will see one of the two possibilities shown on the next page:

116

| | Enantiomers | | Identical | |

Note: the fourth atom or R-group is coming toward you out of the paper.

If you find that you have made the enantiomers, you may change to the "identical" situation merely by exchanging any two gumdrops on one of the models, even the one not on the table. Conversely, if you have made two identical models, you may switch to the "enantiomer" condition by switching any two gum drops (including the one in the air) on one of the models.

To keep up with this kind of thinking you should next construct the glyceraldehyde molecule shown in Figure 16.3 of ECOT. The important thing to note here is that the central carbon atom is bonded to *four different groups*; this is a necessary condition for stereoisomerism. Unless there are four different groups bonded to the central carbon atom, there can be no enantiomers, no stereoisomerism. The four different groups in this case are H, OH, CHO and CH_2OH. Keep the models for the next section, below.

Fischer Projections

The two-dimensional Fischer projections are easily visualized and their usefulness appreciated if you have the model in hand while reading Section 16.6 of ECOT. Figure 16.5 shows glyceraldehyde represented in Fischer projection. Hold the model with the bonds making the "+" shape, vertical and horizontal. Moreover, the CH_2OH group should be *down* and *away*, the CHO group should be *up* and *away*. The other two groups should be coming *toward* you, one to the left and one to the right.

Structures of Sugars

It is time we started using some memory tricks to keep structural facts in mind with regard to sugars. Examine Figure 16.22 of ECOT to see glucose and galactose compared. Variation in the aldohexoses occurs on the side of the ring coming out of the page, particularly on the number-2, -3, and -4 carbons; otherwise, they are the same. For glucose, as you go through 2, 3, and 4, the OH groups are *down* on the number-2, *up* on the number-3, and maybe (see below for the difference) *down* on the number-4. (down/up/down).

For galactose, the only difference is in the number-4 carbon, where the OH is *up*. (down/up/up). Thus we have an easily memorized relationship:

galactose   down/up/up
glucose   down/up/down

Questions Answered

16.1

$$0.08 \times 110 \text{ lb} \left(\frac{454 \text{ g}}{\text{lb}}\right) = 4000 \text{ g of blood}$$

Average concentration of glucose is $\left(\dfrac{0.06 + 0.11}{2}\right) = 0.085 \%$

wt of glucose in blood $= 0.00085 \times 4000 \text{ g} = 3.4 \text{ g}$

117

16.2 $C_{12}(H_2O)_{11}$

16.3 See Table 16.1. a. glucose and fructose; b. glucose and maltose; c. starch and cellulose.

16.4 Since there is a carbonyl group on the number-2 carbon of the 6-carbon fructose, it should be a ketohexose.

16.5 No, the molecule's mirror image could be flipped over and would match the CWXYZ molecule exactly.

16.6 It's easiest to make models of these first.

$$
\begin{array}{ccc}
\text{CH}_3 & & \text{CH}_3 \\
| & & | \\
\text{H}-\!\!\!-\text{CH}_2\text{CH}_3 & \quad & \text{CH}_2\text{CH}_3-\!\!\!-\text{H} \\
| & & | \\
\text{OH} & & \text{OH}
\end{array}
$$

Enantiomers of 2-butanol

$$
\begin{array}{ccc}
\text{CH}_3 & & \text{OH} \\
| & & | \\
\text{H}-\!\!\!-\text{CH}_2\text{CH}_3 & \quad & \text{H}-\!\!\!-\text{CH}_2\text{CH}_3 \\
| & & | \\
\text{OH} & & \text{CH}_3
\end{array}
$$

The Fischer Projection on the right has been rotated 180°.

The two isomers are not superposable.

16.7 Rotate the dropped lens over the fixed one. When the light is completely blocked, rotate the dropped lens back 90° and pop it into the frame.

16.8 A mixture in equal amounts should cancel each other's rotation of light. Racemates are not optically active and are spoken of as being *optically inactive*.

16.9 Add glucose. No, racemates consist of equal amounts of enantiomers of a compound. In the optically inactive mixture, there are not equal amounts of glucose and fructose. More importantly, glucose and fructose are not stereoisomers, so they cannot be enantiomers; they are completely different compounds.

16.10 4; 5; In the noncyclic form, the carbonyl carbon is not chiral, because it does not have 4 different substituents, but in the ring form, it does.

16.11 In cellobiose, the glucose rings are connected by β-linkage; in maltose they are connected by α-linkage.

16.12 Glucose.

16.13 Without lactase the baby cannot effectively hydrolyze the lactose in mild, so he or she will not be able to use lactose as a source of energy. Since babies get about 40% of their energy from lactose, the lactase-deficient baby must be given supplemental nutrition.

16.15 The oxidizing agent will oxidize I^- to I_2 and produce a reddish-brown color that is due to the presence of I_2. A dark blue color will not develop, however. Since cellobiose's β-linkages form long strings that do not curl into a helix, there's no space for the I_2 molecule to interact.

Supplementary Exercises

S1. Which of the following are chiral objects? a. door key; b. dinner fork; c. dinner knife; d. hose nozzle; e. postage stamp.

S2. Suggest the substrate (target molecule) for the following enzymes: a. glucase; b. maltase c. triglycerase; d. cellobiase; e. lactase.

S3. Which of the following solutions would you expect to rotate polarized light? a. table sugar; b. galactose; c. an equimolar mixture of glucose and fructose; d. $BrCH_2OH$;

e.

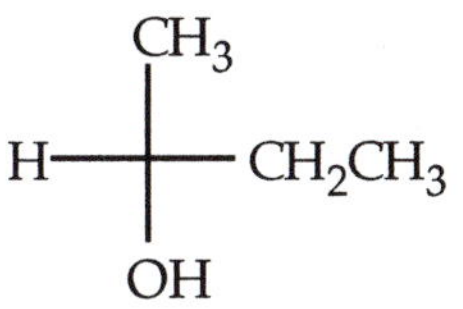

$$
\begin{array}{c}
CH_3 \\
| \\
H-\!\!\!\!-\!\!\!\!-\!\!\!\!-CH_2CH_3 \\
| \\
OH
\end{array}
$$

S4. From the array of molecular structures below, identify any aldoses, ketoses, hexoses and pentoses. Use the combined classification, such as aldopentose, ketohexose, etc., and further classify the monosaccharides shown.

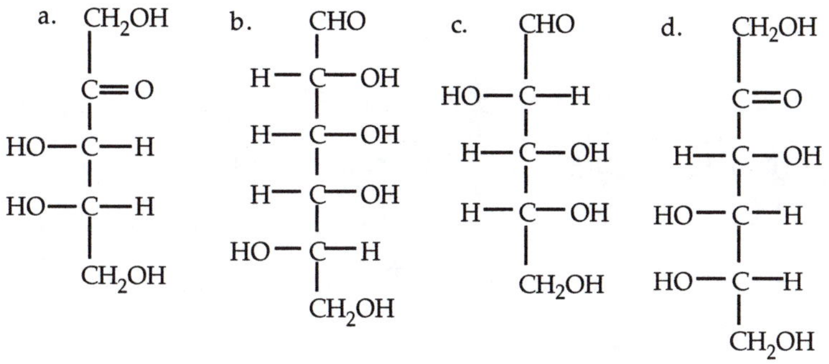

a.
$$
\begin{array}{c}
CH_2OH \\
| \\
C=O \\
| \\
HO-C-H \\
| \\
HO-C-H \\
| \\
CH_2OH
\end{array}
$$

b.
$$
\begin{array}{c}
CHO \\
| \\
H-C-OH \\
| \\
H-C-OH \\
| \\
H-C-OH \\
| \\
HO-C-H \\
| \\
CH_2OH
\end{array}
$$

c.
$$
\begin{array}{c}
CHO \\
| \\
HO-C-H \\
| \\
H-C-OH \\
| \\
H-C-OH \\
| \\
CH_2OH
\end{array}
$$

d.
$$
\begin{array}{c}
CH_2OH \\
| \\
C=O \\
| \\
H-C-OH \\
| \\
HO-C-H \\
| \\
HO-C-H \\
| \\
CH_2OH
\end{array}
$$

S5. How many chiral atoms are there in each of the structures in Exercise S4?

S6. How many possible stereoisomers are there for the molecule in the *a.* structure of S4? In the *d.* structure?

119

S7. How much energy (in Calories) is available from a. 100 g of glucose; b. 100 g of sucrose; c. 100 g of cellobiose; d. 100 g of fructose?

Where You Might Goof Up...

- Being unfamiliar with the meanings of *carbo* and *hydrate*.

- Incorrectly describing the reaction between starch and I_2.

- Incorrectly defining mono-, di-, and polysaccharide.

- Knowing no examples of mono-, di-, and polysaccharides.

- Making mistakes when asked to sketch structures of simple aldehydes and ketones.

- Failing to select aldoses or ketoses from among a list of structural formulas.

- Despite close inspection of molecular models, being unable to identify chiral carbons.

- Given a pair of polarizing sunglasses, being unable to demonstrate their glare reducing properties on common objects while in a typical classroom.

- Being shown a simple molecular model, incorrectly constructing its enantiomer.

- Being unable to explain why an equimolar mixture of glucose (dextrorotary) and fructose (levorotary) still rotates light.

- Not being able to distinguish between α-glucose and β-glucose models.

- Being unable to explain why starch is nutritious for us but cellulose is not.

- Not recognizing the cyclic structures of lactose, sucrose and maltose when shown sketches or models.

- Being unable to predict the number of enantiomers a compound can have after being shown a structural diagram or model.

- Incompletely discussing how cellulose is useful although not energy producing.

- Ineptly drawing that analogy between "lock-and-key" and the way an enzyme acts on a reactant.

CHAPTER 17

Proteins

FIRST AMONG EQUALS

> *Mora rose at daybreak and began preparing the food her family would consume that day. She gathered corn, various types of beans, some special herbs for flavoring. Using a stone she had shaped to match the inner curvature of her favorite ceramic cooking pot, she scraped the inside of the pot to remove the scorched residue of yesterday's meal. In a stone mortar she mashed the corn, beans, and herbs and added some water, mixing until she had a vegetable slurry. She poured the slurry into the cooking pot and put the pot on the coals to simmer for several hours into a thick nourishing soup.*
>
> *She ground some wheat in preparation for making the flat round bread her mother had taught her to make. Grinding wheat was hard work, and as she ground it a few wheat grains fell into the dry dirt; a group of children ran past chasing a running ook-ook bird, and a few wheat grains were kicked into the fire pit and charred.*
>
> *In between her myriad food preparing activities there was time for Mora and her eldest daughter to weave a marriage mat for a relative who was to be married in a few days. There was time for her to make an amulet out of a leather thong and a pretty shiny rock for her youngest child who would be blessed by the shaman during the time of the full moon. And there were just a few moments when Mora could sit back on her heels and gaze at the distant mountain, quiet moments when she could wonder what the other side of the mountain looked like and whether there were people like her beyond the mountain.*
>
> *Without knowing about essential and nonessential proteins and the importance of complementary proteins in her diet, Mora's soup of beans and corn and the bread she made from wheat would have provided her and her family with protein. Supplemented by the nutrition in the green plants she foraged for in season, and by the meat that was available at times, her diet would have been adequate.*
>
> *To be continued in Chapter 18...*

Chapter Overview

* An experiment with egg whites is recommended which, if performed, will aid substantially in the understanding of this chapter.

* What proteins do in our bodies is listed.

* The relationship between amines and amino acids is examined.

* The detailed structure of α-amino acids is presented.

* Essential amino acids, those amino acids humans cannot manufacture, are listed and their dietary sources are identified.

* Complete (high-quality) and incomplete (low-quality) proteins are examined.

* Advice is given to vegetarians so they can stay healthy.

* How amino acids link to form proteins is depicted.

* The primary structure of a protein and its importance is examined.

* Some diseases associated with aberrant or missing proteins are mentioned.

- The hydrogen bond and its involvement in the formation of secondary, tertiary, and quaternary structures of proteins is described, and the egg white experiment is discussed in light of these levels.

For Emphasis

α–Amino Acids

An inspection of Table 17.1 in ECOT reveals that the α-amino acids under discussion (with one exception) can be represented by a single formula containing a carboxyl group (-COOH), an amino group ($-NH_2$), a hydrogen (-H), the α-carbon and an R group; the R group varies, and it is this variation which makes each amino acid different:

$$R-\underset{\underset{NH_2}{|}}{\overset{\overset{H}{|}}{C}}-\overset{\overset{O}{\parallel}}{C}-OH$$

When R = H the amino acid is glycine, when $R = CH_3$ the amino acid is alanine, when $R = CH_2SH$ the amino acid is cysteine, and so forth. The one exception, proline, has no R group as we have described it; because the amine group is part of a ring system attached to the α-carbon, we cannot represent proline with the general formula above.

The simplest a-amino acid to have a stereoisomer is alanine; $R = CH_3$ in this case. Using your gum drop model kit, construct this molecule.

Essential Amino Acids

Table 17.1 of ECOT details the twenty amino acids in human protein. The *essential amino acids* are among those listed; our bodies cannot synthesize them, so they are *essential to the diet*. We must eat foods that contain these unsynthesizeable amino acids to remain healthy, if not stay alive. They always seem to be turning up on quizzes, so you might as well memorize them:

Isoleucine	Lysine
Leucine	Threonine
Phenylalanine	Methionine
Tryptophan	Valine

An effective way of committing a list to memory is by the use of a *mnemonic* (pronounced "ne-mon' ic), a memory aid usually in the form of a sentence, rhyme, or an unforgettable wordplay. (You've probably already memorized one mnemonic. Remember "oil rig" from the redox chapter?) In our examples, the first letter of each word in the mnemonic corresponds to the first letter of the word in the list you are trying to remember. The sillier the sentence or rhyme is, the better it seems to stick. For example:

> In Leningrad, Picasso traced Louise, the mysterious vampire.

The list of amino acids above is in the order of this mnemonic; the *first letter of each word* in the mnemonic is the same as the *first letter of each essential amino acid* in the list.

In	Isoleucine
Leningrad	Leucine
Picasso	Phenylalanine
traced	Tryptophan
Louise	Lysine
the	Threonine
mysterious	Methionine
vampire	Valine

Another one:

Meeting Valine, Lucy tried isolating three Philadelphia lycanthropes.

This mnemonic is particularly good in that, although it does not conform to the order of the list of amino acids above, at least the first two letters of each word are the same as the first two letters of the corresponding amino acid. Take a look:

Meeting	Methionine
Valine	Valine
Lucy	Leucine
tried	Tryptophan
isolating	Isoleucine
three	Threonine
Philadelphia	Phenylalanine
lycanthropes	Lysine

The idea here is to make up your own mnemonic, not to memorize one of these student-generated mnemonics. Your own will stick with you. Try it.

Peptide Bonds

The peptide bond or link in proteins has the structure

$$-\overset{\overset{\textstyle O}{\|}}{C}-\overset{\overset{}{}}{\underset{\underset{\textstyle H}{|}}{N}}-$$

called the *amide* functional group. The link is easy to understand if you draw simple amino acid molecules like alanine and glycine and look at where the water molecule comes from when the amino acids combine to form a dipeptide.

alanine   glycine   alanylglycine, Ala-Gly

Water is formed from the —OH group on the carboxyl group, —COOH, from the alanine molecule and one of the hydrogen atoms from the amino group, —NH_2, on the glycine molecule. We generally write amino acid formulas with the free amino group on the left and the one with the free carboxyl group on the right. As you can see in the dipeptide above, more peptide links can form, since each end of the alanylglycine molecule can form a peptide bond with another amino acid. This can produce a very large protein molecule with a high molecular weight. Such molecules, formed from small molecular units linked together, are called *polymers*. We'll see many more examples of polymers in Chapter 21.

Questions Answered

17.1   Amino acids.

17.2

$$H_2N\text{-}CH_2\text{---}\overset{\displaystyle O}{\underset{\displaystyle \|}{C}}\text{---}O\text{---}H \; + \; HCl \longrightarrow \overset{+}{H_3N}\text{-}CH_2\text{---}\overset{\displaystyle O}{\underset{\displaystyle \|}{C}}\text{---}O\text{---}H \; + \; Cl^-$$

$$H_2N\text{-}CH_2\text{---}\overset{\displaystyle O}{\underset{\displaystyle \|}{C}}\text{---}O\text{---}H \; + \; NaOH \longrightarrow H_2N\cdot CH_2\text{---}\overset{\displaystyle O}{\underset{\displaystyle \|}{C}}\text{---}O\text{---}Na^+ \; + \; H_2O$$

17.3   Only α-aminobutyric acid and β-aminobutyric acid are chiral; γ-aminobutyric acid is not, because no carbons in γ-aminobutyric acid have four different substituents.

17.4   False. Our bodies can synthesize nonessential amino acids, so it is *not essential* that we obtain them from our foods. Both essential and nonessential amino acids are required to build proteins.

17.5   Eggs provide substantial amounts of the essential amino acids in almost the right proportions for humans, so the protein in eggs is of a high quality. Cornmeal has somewhat more leucine per 100 g and does not contain cholesterol.

17.6   Methionine.

17.7   Alanylalanine.

17.8   9 x 3 = 27

$$
\begin{array}{llll}
AAA & AAE & AEA & ATA \\
AAT & AEE & ATE \\
& AET & ATT \\
\end{array}
$$

Repeat with E in the first place and with T in the first place.

17.9   Oxytocin plays a vital role in the release of milk during lactation. Vasopressin, on the other hand, regulates the amount of water retained in the body. The molecules resemble each other in having a ring structure involving a S—S bond between two cysteine molecules. They differ in that in vasopressin the phenyllanine and arginene replace the isoleucine and leucine of oxytocin.

17.10 In sickle cell anemia, valine replaces glutamic acid in one of the chains of hemoglobin. This incorrect substitution causes a small change in the shape of the hemoglobin protein, but alters the entire shape of the red blood cell containing the hemoglobin.

17.11 Globular proteins form small spheres and can move easily in the blood. Fibrous proteins would become twisted and perhaps clog the bloodstream.

Supplementary Exercises

S1.   Sketch chemical structures of each of the following dipeptides: a. Gly-Pro;   b. Leu-Glu.

S2.   Sketch the following tripeptides: a. Ala-Gly-Ala;   b. Cys-Met-Ile

S3.   Devise a mnemonic of your own for memorizing the eight essential amino acids.

S4.   Is there a medicine or drug that can be taken to prevent or cure sickle-cell anemia? If so, what is it? If not, why not?

S5.   Using Fischer projections, sketch D-alanine and L-alanine (Sec. 17.3).

Where You Might Goof Up...

- Forgetting everything you learned about chiral molecules in the last chapter.

- Being unable to list the elements found in proteins.

- Incorrectly constructing models of selected naturally occurring amino acids.

- Being unable to recognize and draw a peptide link.

- Incorrectly listing the essential amino acids.

- Finding yourself at a loss when asked to recommend supplementary foods for an unbalanced or inadequate diet, given charts of relative abundances of protein contained in various foods.

- Being unprepared to distinguish among primary, secondary, tertiary, and quaternary structures on the basis of the hierarchy of complexity.

CHAPTER 18

The Chemistry of Heredity

MOLECULAR GENETICS

Continued from Chapter 17...

Five thousand years after her death Mora's remains were found in North America by twenty-first century archaeologists. Buried beside her were her cooking pot, the grain jar, a dull green stone with a hole in the center through which was threaded a leather thong, a pot scraper, a small piece of woven plant material. Not far from her grave archaeologists found a fire pit. A few tiny slivers of charred ancient wood had survived in the dry climate. Some hard, black nodules turned out to be charred wheat grains.

Mora's skeleton and meager possessions yielded an immense amount of information about her culture. From the wood, the plant material, and a piece of Mora's bone, isotopic dating using C-14 and N-15 (Chapter 5 in ECOT) gave 3000 BCE as an approximate date for Mora's life. An edge of the green stone was carefully scraped, and under the green coat appeared shiny copper. The green outer material was analyzed and found to be a mixture of copper oxides and carbonates, formed when the copper was oxidized in air (Chapter 11). DNA testing on a small piece of Mora's bone showed she was related to people who had lived on the other side of the mountain where a dig had turned up similar remains. This suggested that her ancestors had perhaps formed a number of communities. Anthropologists were interested in what this might say about cultural migrations. Further DNA testing on the residue of scorched material on the edge of the pot scraper and inside the cooking pot gave archaeobotanists insight into the species of corn and beans that grew in the region in 3000 BCE. The charred wheat grains yielded DNA to the molecular archaeologists, as well, helping to build a model of the ecology of Mora's region.

Mora represents the many ancient people whose skeletal and mummified remains have been found since the mid-19th century and who are contributing to our increasing understanding of human history. The exciting new area of molecular archaeology has the potential of providing details of movement patterns, genetic relationships, plant progenitors, and much more, but it's not without controversy. Who owns Mora's remains? The archaeologists who found her and the university they represent? The descendants who claim her as an ancient ancestor? The country in which she was found? Do we have a right to destroy pieces of Mora's bones to obtain the DNA resident there? If we don't glean everything we can from Mora and the items found with her, can we ever put all the pieces of human history together? These are hard questions, but they are becoming more and more important as scientists develop better ways of characterizing DNA in ancient remains.

Chapter Overview

- An experiment to try at home is provided, with questions on the outcome to be answered by molecular genetics.

- A glance at the science of biochemistry is provided by the chemicals in gelatin and pineapple.

- The work of Gregor Mendel is discussed, and phenotype and genotype are introduced.

- Mitosis and meiosis are elucidated.

- DNA is introduced and its relation to the gene is explained.

- The structure of DNA is provided and its amine bases are introduced.

- Messenger and transfer RNA are described.

- The genetic code is elucidated.

- The double helix of DNA is depicted, and the work of Watson, Crick, Wilkins, and Franklin that led to the first model of DNA is clarified.

- The human genome is discussed, and the amazing fact is given that what makes each of us unique is only 0.1% of our DNA.

- Cloning is introduced along with the idea that it takes more than genetics to make two identical individuals.

For Emphasis

DNA — How does it Work?

Why is DNA called the "code of life"? This description is fitting because DNA contains the instructions for building proteins, *all* the proteins in a living organism. DNA controls the makeup of simple bacteria as well as something as complicated as a human being. In Figure 16.12 of ECOT you see a drawing of a section of DNA with the DNA bases hanging off to the side. Although it looks like the bases appear in random order willy-nilly, ECOT pointedly describes the fact that the sequence carries information, intelligence, directions for the synthesis of the primary structure of a protein. Recall that the primary structure of a protein is the *order in which the amino acids* are strung together. DNA tells what this order must be; here is how it does it: The order of the bases tells what the next amino acid in the chain should be. For instance, as the protein chain is being put together, the DNA base sequence adenine-cytosine-adenine (ACA) orders, "Put on a threonine next." The sequence guanine-guanine-guanine says, "Put on a glycine next." The order of three bases in a row (called a *codon*) unambiguously tells what amino acid should be joined to the growing end of the protein chain. Table 16.3 in ECOT shows the codons that represent the specific amino acids.

Several things are to be noted about this table:

- Most of the amino acids can be called up with more than one sequence of bases. This has great survival value for the species as well as the individual. If lysine is to be coded for and there is a mistake in the third member of the codon and "G" appears there rather than "A," lysine will be properly placed nonetheless. A mutation in the third position will not change the message.

- A few amino acids have only one codon.

- The full code, found in most biochemistry books, is complete with stop and start orders.

We realize what a spectacular triumph it is for DNA to transmit properties such as blond hair, large bone-frame, resistance to certain diseases, diminished vulnerability to tooth-decay and myriad other inherited traits through thousands and thousands of years unchanged. DNA is exposed to constant attack by outside forces such as cosmic and nuclear radiation, ingestion of toxins and poisons and exposure to drugs. Unfortunately, when the DNA becomes modified (mutated) it becomes the recipe for all further protein synthesis in the descendants. Disease caused by a genetic defect due to mutation cannot be cured by a dose of medicine. Although research is well underway aimed at giving relief for such diseases as cystic fibrosis, sickle-cell anemia, phenylketonuria and Tay-Sachs disease, among others, success is not yet near.

Questions Answered

18.1   a. reversible; b. irreversible; c. irreversible

18.2   Oxytocin.

18.3   Citric acid.

18.4   Phenotype; genotype.

18.5   No obvious patterns are apparent.

18.6   Deoxyribonucleic acid.

18.7   In biology a gene is piece of a chromosome that carries the information needed to create a specific protein. In chemistry a gene is a piece of the DNA chain with a large number of bases arranged in the right order to direct the formation of a specific polypeptide.

18.8   Both mRNA and tRNA contain ribose as the ring sugar, exist as a single strand, and contain uracil instead of thymine. mRNA contains the information for making proteins and carries this information from the nucleus to the ribosomes in the cytoplasm. tRNA transports amino acids from the cytoplasm to the ribosomes.

18.9   Glycine

18.10 The two DNA strands of a double helix have nucleotides placed in different orders, so the strands cannot be considered identical. However, the nucleotides on one strand match up with the corresponding nucleotides on the other strand by hydrogen bonding. For example, adenines on one strand pair up with thymines on the other and cytosines pair with guanines.

18.11 A genome is the sequence of genes in DNA for a certain species. In other words all of the genes of an individual make up the individual's genome.

18.12 Dolly the sheep; February 1997 in Scotland; There were 277 failed attempts before Dolly.

Supplementary Exercises

S1. In RNA the four amines (guanine, cytosine, adenine, and uracil) can be arranged into sets of three that are called codons. What is the maximum possible number of unique codons that can be formed from these four amines? (Hint: Since codons are read from left to right, UUG is different from GUU.)

S2. If a codon consisted of two amine bases instead of three, how many unique codons could be formed? Is this enough codons to assign an unique codon to each of the amino acids listed in Table 18.1?

S3. In DNA each adenine on one strand of the double helix hydrogen bonds to a thymine on the other strand. The same is true for cytosine and guanine. If the DNA of a newly discovered animal is analyzed and adenine makes up 21% of the amine bases, what are the percentages of thymine, cytosine, and guanine in the DNA?

S4. Use Table 18.1 to list the amino acids in the polypeptide coded for in the mRNA fragment below. Be sure to start at the "initiate" codon and stop at the "terminate" codon.

GGUACGUAUGGAAGUCUGGAAAUUUUAGCUCUAGC

S5. The first adult mammal to be cloned was a sheep named Dolly, but this success happened after 277 failures. What is the success rate expressed as a percentage?

Where You Might Goof Up...

- Being unable to define biochemistry.

- Being unfamiliar with the work of Gregor Mendel.

- Confusing mitosis and meiosis.

- Mixing up the definitions of chromosome and gene.

- Incorrectly listing the three chemical segments of DNA.

- Not remembering what Watson, Crick, Franklin, and Wilkins did.

- Not knowing what a hydrogen bond is.

- Being unable to describe the double helix of DNA.

- Being unprepared to discuss the genetic code.

- Forgetting the roles of messenger RNA and transfer RNA in protein synthesis.

- Being unprepared to discuss the controversy surrounding cloning.

CHAPTER 19

Vitamins, Minerals, and Additives

MICRONUTRINTS AND OTHER FOOD CHEMICALS

Copper is a beautiful, shiny, red-gold metal. Its color is so hard to describe that "copper" is simply the name of the color. It has been used since antiquity to make vessels and jewelry, and one of its major uses today is in electrical wire. Copper's compounds are generally blue or green, the color of the copper(II) ion, Cu^{2+}.

Copper is an essential trace element in a healthy diet, but its function is not well understood, so it's difficult to determine exactly what our daily requirements are. For this reason the National Academy of Sciences doesn't set a Recommended Dietary Allowance for copper, but 3 mg/day of copper are generally thought to be adequate. So how much copper is that?

Reach in your pocket or purse and get a penny. If it's a pre-1983 penny, it weighs about 3 g and is pure copper. If it is dated 1983 or later, it weighs 2.5 g, it has a zinc core, and the copper cladding is so thin that the newer penny is only 2.4% Cu by weight. Let's work with the older penny in our ruminations. It has probably occurred to you that if the penny weighs 3 g and we need 3 mg of copper, we need to eat copper equivalent to a thousandth of an old penny. How much of a penny is that?

Find the letter "o" in "United States <u>of</u> America" on the reverse side (Lincoln is on the <u>obverse</u> side). The "o" is about 1 mm^2 in area. The thickness of the coin on the raised rim is about 1.2 mm, so from the inside of the "o" through the penny to the obverse side is about 1 mm in thickness. So, if you could remove all the copper under the "o" you would have about 1 mm^3 of copper. Copper's density is 9 g/cm^3. So how many mg of Cu are in our 1 mm^3 under the "o"?

$$9 \text{ g/cm}^3 (1 \text{ cm}^3/1000 \text{ mm}^3) = 9 \text{ mg/mm}^3$$

So, we must consume about 1/3 of the volume of Cu under the "o" to get what is thought to be an adequate amount. However, it would be useless for us to try to consume that in the form of elemental Cu. In Chapter 10 of ECOT we saw that the copper half-cell reaction is below the hydrogen half-cell reaction (Table 10.1). This means that elemental copper is not oxidized by the acid in our stomachs. So, do some label reading. What form does copper take in mineral supplements?

Chapter Overview

* A demonstration involving vitamin C is described; it is somewhat like an earlier one involving iodine, except that the tacks are replaced by vitamin C.

* A distinction between macro- and micronutrients is made.

* Trace elements (both major and minor) are listed and a point is made about their necessity for life.

* The distinction between vitamins (both fat- and water-soluble) and minerals is made.

- Some nutritional properties of vitamins A, C, and D are outlined, along with their sources, salutary effects and overdose effects.

- Myths and realities of vitamins are examined.

- The difference in price between "natural" vitamins and synthetic vitamins and their relative worth is argued.

- The interpretation of "health" and "junk" as applied to foods is examined and reading the labels on the box is shown to be a better way of making choices.

- The differences among Recommended *Dietary* Allowances, Reference Daily Intake, Daily reference Value, and Daily Value are pointed out.

- What "additives" means and how they get on the GRAS list is described.

- The use of additives to make food more appealing is covered.

- The use of additives and the activity of EDTA in retarding spoilage is reported.

- The action of stabilizers in making food easier to handle and store is illuminated.

- Life is seen to be a series of tradeoffs when we risk the cancer-causing potential of $NaNO_2$ for its benefits in suppressing botulism.

For Emphasis

This chapter has several sets of information, such as:

- Minerals: What they are, what they do, where they're found in our foods, how much is required;

- Vitamins: What they are, what they do, where they're found in our foods, how much is required;

- The RDA's and the U.S. RDA's: What they are, how they differ, who develops the lists, some examples from each list;

- GRAS list: how additives make the list, some examples of additives on the list, their uses;

- Lists of preservatives: what they are, what they do, how they do it;

- Lists of appeal-enhancers: what they are, what they do, how they do it;

- Lists of stabilizers and other processing aids.

All of these collections of information will be easier to learn if you set up information tables as you have in past chapters. For most of these there are tables in Chapter 19 of ECOT, but you will learn more if you glean the information you need from the chapter material, rather than copying from the tables. After you have collected all you can find in the chapter on a particular subject, compare your information table with the one the author of ECOT has provided.

Questions Answered

19.1 Bones and teeth.

19.2 Iron; anemia.

19.3 Vitamins A and D are fat-soluble; vitamins C and the B complex are water-soluble.

19.4 Carrots contain the precursor to vitamin A which is β-carotene, not the vitamin itself. Our bodies convert the β-carotene into vitamin A.

19.5 The condition is called rickets and results in severely bowed legs and other skeletal deformities.

19.6 Bleeding gums are the first visible symptom. Without vitamin C the collagen in our gums is not replaced when it becomes worn out from the wear caused by eating and brushing. Without the collagen the gums begin to break down and bleeding begins.

19.7 Ascorbic acid is easily oxidized. Since oxidized compounds lose electrons, ascorbic acid is a good source of electrons to be transferred to other reagents.

19.8 Only its molecular structure determines whether or not a molecule is vitamin C.

19.9 Studies show that a well-balanced diet is superior to vitamin supplements when preventing certain diseases. Several studies show that vitamin supplements either cause no reduction in the frequency of certain diseases or actually increase the frequency of those diseases.

19.10 Using what you have learned in ECOT so far, choose an answer and support your opinion with well-reasoned arguments.

19.11 The RDI is the highest RDA for a micronutrient. The RDI for Vitamin D is 10 μg. (See Table 19.3.

19.12 Sodium chloride and calcium phosphate; food seasoning

19.13 Using ECOT's definition of a food additive (not the legal definition), a GRAS substance is a food additive, because ECOT defines a food additive as "anything intentionally added to a food to produce a specific, beneficial result, regardless of its legal status." Under the legal definition of a food additive, however, a GRAS substance is exempted from the Federal Food, Drug, and Cosmetic Act, and therefore is not a food additive.

19.14 Ascorbic acid (vitamin C).

19.15 Amino acids.

19.16 β-carotene.

19.17 Nitrogen and oxygen in EDTA both have nonbonding electrons that they can use to form bonds with other atoms.

19.18 Anticaking agent.

Supplementary Exercises

S1.   Refer to Figure 19.5 in ECOT. How many enantiomers can vitamin C have?

S2.   Teeth are mainly made of hydroxyapatite, $Ca_{10}(PO_4)_6(OH)_2$. A statement in Section 19.1 of ECOT states that it takes about 0.5 g of phosphorus to fix 1 g of calcium in teeth. Can you confirm this from the formula for hydroxyapatite?

S3.   Table 19.1 of ECOT lists dietary minerals; devise a mnemonic to help you select the major minerals from a list of both major and trace minerals.

S4.   a. Given that a 34-oz cucumber contains 37 mg of Ca, how many would a 17-year old have to eat to get half the RDA of that mineral? b. What is the percentage of Ca in that cucumber?

S5.   One cup of cooked spinach contains 14,580 IU of vitamin A. a. If an adult female ingested 1/2 cup of this vegetable, what is the maximum amount of the vitamin that will be unused from this the first day? b. What is the probable fate of this excess?

S6.   A 15-year old male ingests one cup of grapefruit juice (92 mg of vitamin C) and one cup of cooked Brussels sprouts (135 mg of vitamin C) for breakfast.
a.   How much vitamin C is in excess?
b.   How many days will this last him if he eats no more vitamin C-bearing food for a while?
c.   What is the fate of the excess vitamin C?

Where You Might Goof Up...

- Inadequately recounting and explaining the demonstration described at the beginning of the chapter.

- Being unable to distinguish between vitamins and minerals.

- Confusing the RDAs, RDIs, DRVs, and DVs.

- Being unable to justify the different mineral requirements for men and women other than on the basis of body size.

- Failing to advise against leaving a vitamin C drink standing out in the air.

- Omitting solar radiation as one source of vitamin A.

- Being unable to perform calculations involving REs, IUs and grams when dealing with amounts of vitamins.

- Having no explanation for the fact that we eat carrots for vitamin A, but the carrots contain none.

- Inadequately defending the thesis that natural vitamins are the same as synthetic vitamins.

- Being unable to go beyond the superficial labeling on a package when evaluating the worth of the contents as food.

- Forgetting what GRAS stands for.

- Not being able to list at least one antioxidant, one microorganism inhibitor, one humectant, one emulsifier, and one anticaking agent, and give an explanation of how each works.

- Suggesting that EDTA reacts with vitamins rather than some minerals.

- Writing less that 150 words on the trade-off between the upside and the downside of $NaNO_2$ as an additive.

- Maintaining that vitamin C acts only as a vitamin when added to food.

- Being unable to discuss "additive" from the standpoint of the legal definition and compare it to ECOT's practical definition.

CHAPTER 20

Poisons, Toxins, Hazards and Risks

WHAT'S SAFE AND WHAT ISN'T

> *The fugu fish lives in the waters off the islands of Japan and participates in a survival mechanism for the school but not himself. This fish concentrates a lethal agent, <u>tetrodotoxin</u>, in the liver. A fugu's predator has one last meal. Curiously, selected organs of the fugu are considered a delicacy in Japanese restaurants, so the important first step in cleaning the fugu is to remove the liver intact lest it rupture and release the toxin into the rest of the fish. This operation is so important that chefs must be licensed by the government for this delicate procedure. Their licenses are proudly worn.*
>
> *Yoshio Ikeda, plying his fisherman's trade, found a fugu in his net and decided to enjoy the delicacy without the high tariff charged in the restaurants. He knew that the liver was the first target of dissection, but in the choppy sea his knife slipped a little, so little that it did not seem important. The last thing Yoshio heard was the clicking of his chopsticks on the deck of the trawler.*

Chapter Overview

- An experiment is described, contaminating the environment, that almost everyone has done already. Skip this one unless it's really necessary.

- The difference between *poison* and *toxin* is explained.

- Lethal events are described involving substances we consider safe.

- A reasonable method for measuring the strength of a poison is detailed; calculations are made.

- A list of the most lethal substances is made and their LD_{50}s are discussed.

- The relative strengths of natural poisons versus those the chemist can concoct are compared. Nature wins.

- A definition of safety is introduced that includes several aspects not usually considered.

- The Government's role in deciding what is safe, from food additives to the packaging of medicine, is described, and the ADI is defined.

- The Delaney Amendment as exception to regulation is recounted.

- Saccharin's ups and downs are discussed, and the safety of eating a mango is argued.

- How vegetables make their own insecticides (which we eat) is reported.

- The ethical trade-off of killing little animals to save big ones is discussed.

- It is asserted that there is no such thing as absolute safety.

For Emphasis

There are only two mathematical relationships to be understood to deal with quantitative aspects of this chapter: LD_{50} and ADI.

Lethal Dose for 50%

The LD_{50} is arrived at in the following way: A large quantity of living specimens is exposed to increasing concentrations of the toxin in question. The dosage is raised geometrically, using µg quantities. (µg is the abbreviation for *micro*gram; $1 \text{ µg} = 1 \times 10^{-6}$ g. So, $1 \text{ g} = 1,000,000$ µg, and there are 1000 µg in 1 mg.) The quantities used might be 1 µg/day, 2 µg/day, 4 µg/day, 8 µg/day, etc. and the cumulative mortality in the population resulting from each increase in dose is recorded. The data are plotted and inspection of the graph reveals the dose at which 50% of the population has been killed. Since average weight of animals varies, the data are reduced to the *per kilogram* value for comparison and uniformity. To extend these data to human beings, it is necessary only to multiply the LD_{50} value by the weight of the person to find the dose which would be lethal to half of the human population the person represents.

Example 1: The widely used herbicide Propanil has an LD_{50} of 1.38 g/kg orally in rats. What dosage would be lethal to 50% of a human population represented by a 190 lb (86 kg) person?

> **Solution**:
> $$86 \text{ kg}\left(\frac{1.38 \text{ g}}{\text{kg}}\right) = 120 \text{ g of Propanil}$$

Example 2: The alkaloid senecionine has been determined to have a lethal effect on half the mice (3.8 g average weight) in a large population at the 0.25 mg level. Calculate the LD_{50} for this compound.

> **Solution**:
>
> Since the lethality data are always reported for a *one kilogram* specimen, we convert the 0.25 mg to the kilogram dosage:
>
> $$0.25 \text{ mg}\left(\frac{1}{3.8 \text{ g}}\right)\left(\frac{1000 \text{ g}}{\text{kg}}\right) = 66 \text{ mg/kg}$$

ADI: Acceptable Daily Intake

The second quantity of interest is the ADI, the Acceptable Daily Intake, a measure applied to additives. The arithmetic associated with this measure comes about from the definition of the ADI: a safety factor of 100, or 1% of the maximum amount of chemical additive which produces no harmful effect. The ADI is determined by feeding at least two different test species increasing amounts of proposed additive until effects of the additive are observed. The largest amount of additive which produces *no* observable effect is noted and defined as the no-effect level. Usually, 1% of this amount is allowed as the daily intake for commercially available foods. The word "usually" indicates that there are exceptions to the above statement. The animals are tested over longer periods of time than the LD_{50} determination experiments. The animals are examined for such effects as reduction in potential for propagation, damage to progeny or incidence of cancer; if animals are positive for any of these conditions, the additive may be banned outright.

Example 3: What is the ADI of an additive whose no-effect quantity is 42 mg/kg per day?

> **Solution**:
>
> Multiply the no-effect amount by 0.01:
>
> 42 mg/kg x 0.01 = 0.42 mg/kg per day

Pesticides, Natural and Otherwise

Plants, both ones we eat and others, are very efficient generators of poisons or toxins. Tomatoes and potatoes are very good at self-defense: they both generate poisons that are lethal to humans. Luckily for us the poisons reside in the leaves of the plants, rather than the edible parts. Many other plants produce poisons to protect themselves against attack from the outside; from apples to zucchini, with Brussels sprouts, comfrey tea and tarragon in between. Some of these active agents in the plants have been identified, but have not had their carcinogenic potentials assessed. Estimates suggest that 50% of them may be carcinogenic.

Consideration of a point made in the *Perspective* section of the current chapter of ECOT, namely, that there are toxins, poisons and carcinogens to be found naturally in our foods shows how an informed public might avoid costly mistakes. Some years ago there was much publicity attendant on a synthetic agent, ALAR, used to retard the ripening of apples. One of the breakdown products of ALAR is an amine known to be carcinogenic. Millions of dollars worth of apples were withdrawn from the market after attention was drawn to this fact in the press. Small apple growers underwent serious financial difficulties, school children were deprived of balanced lunches and other inimical results flowed from this publicity. What was not emphasized by the press was the fact that one mushroom naturally contains ten times as much of the same amine as a serving of juice pressed from two apples treated with ALAR.

Questions Answered

20.1   a. Consider a poison as something we may eat, drink, breathe or otherwise ingest: strychnine, sodium cyanide, hydrogen cyanide (a gas), lead, mercury, or arsenic. Note that these are mostly inorganic compounds. Reading the labels of substances stored under the typical kitchen sink will help flesh out these lists.
b. A toxin is usually thought of as coming from plant or animal origin and extracted for use or misuse: curare, nitrobenzene (found in shoe polish although absorbable through skin), a third helping of rhubarb pie. (Remember: "The toxin is in the dosage.")

20.2   Decreased the death rate; 46 per year; 25 per year.

20.3   The animal species and the method of administration.

20 4   Home-preserved foods; Heat the foods at temperatures high enough and for the appropriate length of time to kill the microorganisms.

20.5   Besides being toxic, isoflurophate, parathion, and sodium cyanide have other properties that make them useful. Isoflurophate is used to contract the pupil of the eye in veterinary medicine; parathion is an effective insecticide; and sodium cyanide is used in the metallurgy industry to extract gold from its ores. Compounds can have a variety of properties, so even though a compound may be extremely hazardous in one situation, it could be absolutely essential in another.

20.6 Aflatoxins in moldy peanuts are toxic and can cause cancer. Solanine in potatoes containing fresh sprouts inhibits nerve impulses and can lead to death in humans and other species.

20.7 Thalidomide is hazardous to developing fetuses in pregnant women when used as a sleeping pill. When used to treat mouth ulcers, rheumatoid arthritis, leprosy, AIDS, tuberculosis, and transplant rejection in non-pregnant individuals, thalidomide's benefits outweigh its hazards.

20.8 a. FDA; b. EPA; c. OSHA

20.9 ADI is the acceptable daily intake of a food additive. It is determined by calculating 1% of the maximum daily amount of additive that produces no observable effect on laboratory animals.

20.10 The Delaney Amendment takes effect when a food additive is found to cause cancer in any animal at any concentration.

20.11 Public pressure on Congress led to a moratorium against FDA action to ban saccharin. Essentially, Congress prevents the FDA from using the Delaney Amendment to ban saccharin.

20.12 Ethyl acetate.

20.13 a. Myristicin may cause hallucinations; b. Glycyrrhizic acid may cause hypertension and cardiovascular damage; c. Oxalic acid may cause kidney damage.; d. Tannic acid and other tannins may cause mouth and throat cancer; e. Allyl isothiocyanate may cause tumors.

Supplementary Exercises

S1. Tetrodotoxin has an intraperitoneal (injection into the abdominal cavity) LD_{50} of 10μg/kg in mice. If this were the same as oral ingestion, where would it fit in ECOT's Table 20.2?

S2. Some commercially available acetaminophen contain 300 mg of the active ingredient per tablet. a. How many tablets would an 80 kg person have to ingest to exceed the LD_{50} for this medicine?

b. If this person could take 3 tablets with each 200 mL gulp of water, how much water would be drunk to consume this many tablets?

S3. What is the ADI of a substance for which the maximum daily amount that produces no observable effect on laboratory animals is 0.50 mg/kg of body weight?

S4. In determining the lethality of brucine, an additive for lubricants, it was found that half a population of 250-g rats was killed when fed this alkaloid at the 2.5×10^{-4} g level.
a. Calculate the oral LD_{50} in rats in mg/kg.

b. Where does brucine fit in Table 20.2 of ECOT?

c. Calculate the size of the human population, half of which would be killed by 1 g of brucine. Assume 80 kg humans.

Where You Might Goof Up...

- Not knowing the difference between a toxin and a poison.

- Expecting that poison- or toxin-free foodstuffs can be reasonably made available.

- Being unable to list incidents where substances commonly eaten or drunk have poisoned and even killed those who ingested inordinately high amounts.

- Given appropriate data, calculating the LD_{50} of a substance incorrectly.

- Given the LD_{50} for a substance, calculating incorrectly the amount needed to kill 50% of a representative mass-defined population.

- Distinguishing incorrectly between LD_{50} and ADI.

- Given appropriate data, incorrectly calculating the ADI for a substance.

- Being unable to list three reasons for overriding quantitative data in ADI determinations.

- Being unable to write a short essay on the subject of naturally occurring toxins in the fruits and vegetables commonly available in grocery stores.

- Being unable to discuss safety as *the acceptability of risk*.

- Being unable to list, besides the FDA, some other agencies and their jurisdictions that oversee health and safety.

- Giving a poor explanation of the Ames test.

CHAPTER 21

Polymers
and Plastics

THE PLASTIC AGE

In an attempt to trip up King Claudius who he is certain engineered the death of his father and usurped the throne, Hamlet, Prince of Denmark, has a troupe of traveling actors reenact the details of his father's death, details Hamlet learned from his father's ghost. His plan is to so unnerve Claudius that he shows his guilt in the old king's murder. "The play's the thing," declares Hamlet, "wherein we'll catch the conscience of the king." (Act II, scene ii)

A modern landfill engineer, unconcerned with palace intrigue but aware of Shakespeare's love of plays on words, might honor the Bard by saying "the clay's the thing wherein we'll catch polluting things!" Clay is a silicate,* an inorganic polymer containing the two-dimensional network of anionic SiO_4 tetrahedrons described in ECOT. In clays these silicate sheets alternate with sheets of aluminum oxide, Al_2O_3, in which water molecules are imbedded. Dangling hydroxyls (-OH) provide polar links (Sections 3.14 and 16.11) between the layers and make clays very hydrophilic. They not only attract water molecules, but other hydrophilic substances as well can be trapped between the layers of the clay.

Because of their ability to trap water, clays have been seen as almost perfect materials to contain hazardous waste in landfills. Pollutants move through the soil carried principally by water (leaching); the trapping of water by the clay layer slows the leaching process. But like the poisoned rapier that killed Hamlet at the end of the play, clay's affinity for water is a double-edged sword where toxic waste is concerned. Recent research has shown that clay's hydrophilic character keeps hydrophobic organic solvents from adhering to the layers, and the solvents diffuse much faster than expected. The result is that landfills thought perfectly safe 15 to 20 years ago are now cleanup problems.

But there's hope. A new technology permanently immobilizes contaminants by injecting a mixture of reactants to render the pollutants harmless and cement to harden the products in place. Another radical approach injects layers of water and clay to make a hazardous waste "lasagna" which is then electrolyzed (ECOT Chapter 11) to draw hazardous compounds out of the soils (see "Further Reading").

Clays have played a part in civilization from its beginning. From Mora's ceramic cooking pot used in the time before recorded time (Chapter 17 in this Guide) to fine 19th century European porcelain to the cleanup of the offal of 20th century civilization, this inorganic polymer has lent its chemistry to human use. Research continues, because a landfill carelessly left to leach pollutants to the environment becomes a later generation's problem. Like Shakespeare's King Lear, we work to solve the problems of the present, so "that future strife may be prevented now." (King Lear, Act I, scene i)

Chapter Overview

• An immensely practical demonstration that addresses waste management problems is described.

• The difference between plastics (macroscopic) and polymers (molecular) is related and examples given.

• Condensation and addition polymerization are defined, examples given and equations shown.

---

*Might Petruchio have used this word to describe Katherine's refusal to be straightforward with him in their repartee in The Taming of the Shrew, Act II, scene i?

- The thermal, chemical and mechanical properties of plastics/polymers from Bakelite to vinyl are discussed.

- The serendipitous adventures of Nobelists and other chemists are narrated.

- The structure of nylon is elucidated

- Condensation and addition polymers are discussed.

- An important inorganic polymer, silica, is described.

- One desirable attribute of plastic, its great chemical stability, is also a liability: it makes the plastic last, and last, and last . . . .

For Emphasis

This chapter begs to be organized into a very large information table. Get a piece of poster board or other large paper and start on the left with the basic information of Table 21.1. To be complete, add the silicates to this list. Then go through the chapter and pick out other column headings, such as *monomeric units, polymer category (homopolymer/copolymer), mode of polymerization, thermal properties, etc.*

On the back of the poster put examples of the various modes of polymerization. Write the equations of condensation and addition where they apply to the various polymers.

Questions Answered

21.1 Plastics have low densities, so they have relatively large volumes when compared to their masses. That is why plastics make up a disproportionate share of the trash volume.

21.2 Starch, cellulose, protein.

21.3 Condensation: nylon, starch, protein, cellulose; addition: polyethylene.

21.4 a. homopolymer; b. copolymer; c. copolymer; d. homopolymer; e. homopolymer; f. homopolymer

21.5 No, polyethylene does not contain double bonds, so it cannot have *cis* and *trans* geometric isomers.

21.6 a. chlorine; b. sulfur.

21.7 Cellulose, nitric acid, and cellulose.

21.8 Methylene.

21.9 a. homopolymer: caprolactam; b. copolymer: adipic acid and 1,6-diaminohexane.

21.10 No. Polymers like polyesters have repeating units of monomers that form long chains. Triglycerides do not contain monomers and do not combine to form long chains.

21.11 Polystyrene foam is lightweight, a good shock absorber, and a good thermal insulator.

21.12 a. Oxygen: polyvinylacetate, polymethylmethacrylate (Lucite, Plexiglas) b. Chlorine: polyvinylchloride, polyvinylidine chloride (Saran); c. Nitrogen: polyacrylonitrile (Orlon, Acrilan, Creslan) d. Fluorine: teflon

21.13 Producing plastic trash bags.

21.14 If we consider all polymers being formed from ethylene or a substituted ethylene, those formed from an ethylene having asymmetric substituents on one of the carbons can form atactic or syndiotactic polymers. In Table 21.2 all but ethylene, vinylidine chloride and TFE can qualify.

21.15 For soft drink bottles, PET is preferred because gases can't pass through it easily, so the soft drink won't lose its carbonation quickly. Polypropylene is preferred for larger containers because it is stronger than PET.

Supplementary Exercises

S1.   Using a dash to represent a shared pair of electrons, show how a network of Si and O atoms can be linked so that the simplest formula is SiO_2.

S2.   Nylon 46 is formed by the reaction of 1,4-diaminobutane with adipic acid. Draw the structure of Nylon 46.

S3. Polypropylene has the structure

$$\left[\begin{array}{cc} H & H \\ | & | \\ -C & -C- \\ | & | \\ H & CH_3 \end{array}\right]_n$$

a. Write the formula for the monomer of polypropylene.

b. What kind of polymer is polypropylene?

S4. Examine the bottoms of some plastic containers, and put the numbers that characterize the various plastics on your information sheet.

S5. Visit a recycling center and write a short essay on the method used to separate unwanted plastics and other trash from the various classes of polymers.

Where You Might Goof Up...

- Being unable to distinguish between a plastic and a polymer.

- Being unable to pull out the repeating unit of a polymer, given a structural diagram of the chain.

- Improperly labeling a group of structural diagrams representing, say, polyethylene, nylon, cellulose and Bakelite.

- Improperly describing vulcanization and its effects on the properties of natural rubber.

- Forgetting the names and properties of the most important inorganic polymers on this planet.

- Running out of facts and ideas before reaching 250 words for an essay on the plastics disposal problem.

CHAPTER 22

Cosmetics and Personal Care

LOOKING GOOD AND SMELLING NICE WITH CHEMISTRY

Continued from Chapter 16 . . .

The young king looked into the polished silver mirror and grunted impatiently. "Huh, Na-sept-tu-nu. Aren't you finished yet?" The king's barber smiled and bowed. "In a moment, sire. You must look your best to inaugurate the feast of Geb." With a razor-sharp bronze (Cu and Sn alloy) knife Na-sept-tu-nu shaved away the last of the king's black hair. "There," he said, "ready for the royal wig. But first, the eyes."

On the king's dressing table were the tools, paints, and ointments of the barber's trade. Beautifully carved jars of translucent alabaster ($CaSO_4 \cdot 2H_2O$) held eye paint. One jar was the king's favorite; on its top was a carved white cat with its bright pink tongue stuck out, its front paws crossed casually, eyes fixed on its beholder. It reminded the young king of his favorite white cat from his childhood. He had named her Sekh-met, and she had slept at his feet. Next to the cat jar was a carved ebony box. It held kohl, a dark gray powder (antimony trisulfide, Sb_2S_3), which was mixed with castor oil and painted around the king's eyes. Eye paint was a fashion with a purpose: it provided some protection from the reflected light of the harsh desert sun.

Na-sept-tu-nu poured castor oil onto a polished green schist (metamorphic rock) palette and added a small spoonful of kohl. With a ceramic pestle he gently ground the oil and kohl together, until a dark gray paint resulted. Using an ivory kohl stick he outlined the king's eyes. Then, spooning some green powdered malachite [$Cu_2CO_3(OH)_2$, mined in the eastern desert] from a carved wooden box, he made an oil paste and added a thin green line to outline the black line he had already applied. To protect the young king's lips from drying in the sun, the barber painted them with oil containing rouge (iron oxide, Fe_2O_3).

From a little squat alabaster jar Na-sept-tu-nu scooped a dollop of unguent perfumed with fragrance pressed from lotus flowers. He massaged the salve into the king's neck and shoulders. "Now you're ready, Sire. It's time to dress."

"Look, it's Pharaoh!" a child shouted. Holding a shallow wine bowl aloft for the priest's blessing, Tut-ankh-amun signaled the beginning of the feast of Geb, god of the Earth.

Chapter Overview

• The complexity with which cosmetics and toiletries are intertwined with chemistry is narrated.

• The categories of cosmetics and toiletries are listed.

• The utility of surfactants in shampoos and toothpaste is described.

• The structure of hair and how surfactants are engineered to do (and not overdo) the job on it are portrayed.

• The chemistry of permanent waves is covered in great detail.

• The compounding of toothpaste is reviewed.

- The structure of skin and the way lotions, creams, antiperspirants, deodorants, and colorants work on it are illuminated.

- The three notes of perfume are listed along with some of the compounds which emit them.

- Electromagnetic radiation is discussed, and suntans, sun block, vitamin D, and skin cancer enter into another trade-off discussion.

For Emphasis

Although there is much new information in this chapter of ECOT, it is also a review of topics in earlier chapters. In the table below are a few questions to guide you in quizzing this chapter and pulling together many of the concepts you have learned in this course.

Chapter	Name	Quizzing This Chapter
6	Introduction to Organic Chemistry	Why are functional groups essential to this chapter?
10	Acids and Bases	Why is pH emphasized in hair-care products and toothpaste?
11	Oxidation-Reduction	What is another way to define oxidation?
13	Surfactants	Why are surfactants important in shampoos and toothpastes? What surfactants are used and why?
15	Fats and Oils	Why is myristic acid, a fatty acid, included in many shampoos?
17	Proteins	What is keratin, and what are the three types of bonds that hold it together? How does a permanent change protein structure?
20	Poisons, Toxins, Hazards and Risks	What are the risks of hair dyeing? How are new dyes approved for use in cosmetics?

Now quiz the chapter yourself and make a table of your own.

Questions Answered

22.1 A cosmetic is anything intended to be applied directly to the human body for "cleansing, beautifying, promoting attractiveness, or altering the appearance." In ECOT a cosmetic is a substance applied directly to the human body to make it more attractive.

22.2 Synthetic detergents and other surfactants .

22.3 The primary quality is the suppression or elimination of perspiration; secondary qualities include convenient to use, won't stain clothes, adheres to skin but can be removed by bathing, has a pleasant odor or none at all.

22.4 They are more soluble in cold water than the sodium salt and they don't dry out hair as much as the sodium salt does.

22.5 Sulfur.

22.6 The dye molecules were assembled inside the cortex from other molecules that could penetrate into the cortex

22.7 Sulfur.

22.8 Plaque removal.

22.9 From top to bottom: stratum corneum (25 to 30 layers of dead cells), epidermis (upper layer of the skin with a layer of actively dividing cells), dermis (lower layer of the skin containing nerves, blood vessels, and sweat glands)

22.10 Antiperspirants contain aluminum compounds that inhibit sweating by closing the ducts of the eccrine glands and kill odor-causing bacteria; deodorants mask odors with fragrant ingredients and contain antibiotics that kill the bacteria that generate offensive odors.

22.11 The lengthening mascara makes the eyelash ends thicker and more noticeable.

22.12 The top note is the first scent noticed. (phenylacetaldehyde). The middle note is the most noticeable scent. (2-phenylethanol). The end note is the lingering scent. (civetone).

22.13 Frequency.

22.14 A sunscreen blocks UV-A and UV-B light, while a tanning lotion blocks only UV-B. PABA (*p*-aminobenzoic acid) is used in tanning lotions, and benzophenone is used in sunscreens.

Supplementary Exercises

S1. Inspect the structure in Figure 22.8 of ECOT and explain why FD&C Blue No. 1 would be expected to wash out of hair easily.

S2. Inspect the structures in Figure 22.9 of ECOT and explain why they would be expected to last through several washings. (Hint: Compare their structures with FD&C Blue No. 1.)

S3. Some frugal people wash their hair in a very mild soap and use vinegar or diluted lemon juice as a final rinse, believing that the results are as good as with an expensive shampoo. Refer to Figure 22.6 of ECOT and give a justification for this.

S4. The LD_{50} for thioglycolic acid in rats is 0.15 mL/kg and its density is 1.52 g/mL. How many mL would 100 52-kg individuals have to ingest (accidentally!) for half of them to die?

S6. Based on your knowledge of the effects of chemical structure and mass on vapor pressure, explain why phenylacetaldehyde is a "top note," whereas civetone is a "low note."

S7. Two hours in the sun with an SPF 12 sunblock is equivalent to how many hours in the sun without any skin protection?

Where You Might Goof Up...

- Failing to distinguish between a shampoo and a toothpaste on the basis of the desired attributes of each.

- Being unable to distinguish among temporary, semipermanent and permanent hair dyes on the basis of chemical structure.

- Being unable to describe the step-by-step process of a permanent wave at the molecular level.

- Being unable to outline the three notes of a perfume and describe how they are different aesthetically and chemically.

- Confusing frequency and wavelength when discussing light.

- Relating erroneously energy and frequency of light.

- Interpreting the sunscreen protection factor (SPF) mistakenly.

- Incompletely discussing the health liability of a gorgeous suntan.

- Failing to pull together all the chemistry concepts that are reviewed in this chapter.

CHAPTER 23

Medicines and Drugs

CHEMICALS AND THE MIND

Idnam Gursel tends his farm on the southern edge of Lake Van, close to the border with Iran. He grows wheat and poppies, he has a few goats and three sons. Although his wife might rub some of the poppy sap on the gums of a teething child, Idnam does not partake of the joys of that plant. He has enough enemies to the East in the Kurds and the Persians; although he can read only enough to recite a few prayers, he knows that he does not need another enemy inside. He has heard of the infidels to the West who stand in the shadows of buildings, eyes and noses running, bent with cramps, shaking — but with money in their hands. Maybe a little less wheat next year, a little more poppy.

Chapter Overview

- The student's perceptivity is tested to introduce the idea of misperception and illusion.

- Analgesic, antipyretic, anti-inflammatory, narcotic, vasoconstrictor, endorphin and placebo are defined.

- Over-the-counter medicines/drugs like aspirin and their substitutes are described.

- Narcotics, including a heroic effort against addiction gone sour, are recounted.

- The addictive and lethal properties of the alkaloids nicotine and caffeine are described.

- Drugs and medicines are differentiated.

- The arbitrariness of "natural" *vs.* "synthetic" medicines is exposed.

- The ambivalence of our society towards abusable drugs is addressed.

- The ideal drug is defined, but its development deemed problematical.

- The hallucinogens, sacerdotal and otherwise, are described.

- Tranquilizers and anti-depressants are examined.

- An obsessing question: "Why should the *animal* brain have evolved with receptors within them that only molecules produced by *plants* can fit?"

For Emphasis

This chapter lends itself well to an information table. Headings might include *Analgesics, Narcotics*, etc. Give examples of each class of compound. Include in your table the structure(s) that characterizes each of the classes you choose.

A second information table may give you another way to look at the compounds in this chapter (and to make some sense of them). List down the page all the compounds that have structural formulas given in this chapter. Across the top you might have headings *Similar to... , Functional groups, Contains the structure of..., Physiological activity*, etc. Take scopolamine, for example (Fig. 23.12). It is similar to atropine and cocaine in structure. It has a phenyl group, a substituted amine, a carbonyl group, or, when you look more carefully, you see that the carbonyl group is part of an ester. Scopolamine contains the structures of tropane, benzene, etc. Looking at each of these complicated structures in terms of the parts that make it up helps to simplify the learning process.

Another aid to the learning process is the use of flash cards containing the structure of each compound discussed. A variation on the flashcards is to photocopy structures that are obviously similar, like morphine, codeine (both in Fig 23.10), levorphanol, oxycodone, and oxymorphone (all in Fig. 23.15). Cut out the photocopied molecules and paste them on a single piece of paper. Recopy the paste-up onto Mylar [poly(ethylene terephthalate), Sec. 21.10, available as overhead transparency material], cut the molecules out in the same size squares, and overlay them. Hold the stack on one side, and fan through them with the thumb of your other hand. The differences will be obvious. Then compare one with another by overlaying them. Write a description of each molecule and its differences from the others.

Questions Answered

23.1   Acetic anhydride adds the acetyl group to salicylic acid.

23.2   The "spir" part of "aspirin" comes from the acid derived from the botanical genus of the meadowsweet plant.

23.3   A true buffer contains a weak acid and one of its salts (which is basic). "Buffered" aspirin does not contain a salt of acetysalicylic acid, but contains other bases.

23.4   Two benefits of aspirin are that it is an analgesic (reduces pain) and is an antipyretic (reduces fever). Two risks are that it can cause nausea and Reye's syndrome in children and adolescents.

23.5   Liver damage.

23.6   A good "Think, Speculate . . ." question!

23.7   One of the hydroxy groups of morphine has been methylated to produce codeine.

23.8   Heroin is a much more powerful narcotic than morphine and smaller doses would be needed to produce the same effect.

23.9   Tropane alkaloids contain an ester group as well.

23.10 Carbon monoxide.

23.11 Another good "Think, Speculate . . ." question!

23.12 Phenol (c) and aromatic (e).

23.13 The hydroxy group of oxymorphone has been methylated to produce oxycodone.

23.14 The carbonyl group.

23.15 They both contain amine groups ($-NR_2$). LSD also contains a carbonyl group($C=O$).

23.16 a. β; b. α.

23.17 Fluorine.

23.18 Yes; compare with formulas in Fig. 23.18 in ECOT.

23.19 Glycine, phenylalanine and tyrosine.

23.20 a. tyrosine; b. glutamine.

23.21 It is harmless and considered to be ineffective in treating a particular illness.

Supplementary Exercises

S1. The great sensitivity of the animal organism to small changes in molecular structure is demonstrated by comparing atropine (LD_{50} = 75 mg/kg) to scopolamine (LD_{50} = 3800 mg/kg). What single structural difference exists between these two molecules? (Hint: See Fig. 23.12 in ECOT.)

S2. What does the suffix "-aine" imply about compounds studied in this chapter of ECOT?

S3. Go back to Chapter 1 of this *Guide*, and review the steps of the scientific method. Write a short essay about how a drug company uses the scientific method when it tests a new medication on a human test group using a double-blind study.

S4. From this chapter, write the names of two compounds that contain each of the following functional groups: a. benzene rings; b. amine groups; c. carbonyl groups; d. methyl group; e. hydroxyl group.

S5. In S4 above, name the function of each compound you chose. For example, if one of your compounds is β-phenylethylamine, it functions as a *stimulant*.

S6. Give examples from this chapter of medicines and drugs that come from (or originally came from) plants. In each case, name the plant.

Where You Might Goof Up...

- Being unable to list the addictive drugs mentioned in this chapter of ECOT.

- Describing the heroic properties of heroin without mentioning its downside.

- Being unable to explain why a vasoconstrictor is sometimes compounded with injected anesthetic.

- Overlooking the place of Naloxone in any discussion of the action of opiates on humans.

- Improperly designing a double-blind experiment.

- Failing to distinguish between endorphins and enkephalins.

- Being unable to justify the withholding of aspirin from children in the face of complaints of pain or discomfort.

- Failing to distinguish among antipyretic, analgesic and anti-inflammatory.

- Being overly impressed by "buffering" in aspirin tablets.

- Inadequately discussing the properties of the ideal drug.

Answers to Supplementary Exercises

Chapter 1

S1. a. By now you should see that a pattern has emerged in the way we set up these problems. We want to calculate the number of grams of lithium, so we set up the ratio in the form of lithium/fluorine. This tells us how many g of lithium are needed per g of fluorine.

b. Multiply this number by 12 g of fluorine.

$$\left(\frac{6.9 \text{ g of lithium}}{19.0 \text{ g of fluorine}} \right) 12 \text{ g of fluorine} = 4.4 \text{ g of lithium}$$

S2. a. This problem is exactly like Exercise 14 on p. 14 in ECOT. Since the question asks for a *qualitative* response, think it through as you did Exercise 14.

b. $\left(\dfrac{79.9 \text{ g of bromine}}{39.1 \text{ g of potassium}} \right) 10.0 \text{ g of potassium} = 20.4 \text{ g of bromine}$

S3. Since the problem states that sulfur remains unreacted, sulfur must be in excess, so we will find out how much of it is actually used when 15.0 g of calcium react.

$$\left(\frac{32.1 \text{ g of sulfur}}{40.1 \text{ g of calcium}} \right) 15.0 \text{ g of calcium} = 12.0 \text{ g of sulfur}$$

Since we started with 15.0 g of sulfur, 3.0 g of sulfur are left after the reaction.

S4. There are two approaches that can be used to solve this problem.

Method I:

Talk yourself *intuitively* through the problem. 40.0 g of calcium requires not quite twice as many grams of chlorine, or 70.0 g. We've been given 50.0 g of calcium and exactly twice as many grams of chlorine. But the 50.0 g of calcium will use up not quite twice as many grams of chlorine, so the chlorine will be in excess. Now to find out how much chlorine is used, set up the usual equation using the ratio.

$$\left(\frac{70.0 \text{ g of chlorine}}{40.1 \text{ g of calcium}} \right) 50.0 \text{ g of calcium} = 87.3 \text{ g of chlorine}$$

The quantity of chlorine in excess is 100.0 g − 87.3 = 12.7 g.

Method II:

Set up the two reciprocal ratios and calculate how much of each element is required to react with the given amount of the other element.

$$\left(\frac{70.0 \text{ g of chlorine}}{40.1 \text{ g of calcium}}\right) 50.0 \text{ g of calcium} = 87.3 \text{ g of chlorine}$$

and

$$\left(\frac{40.1 \text{ g of calcium}}{70.0 \text{ g of chlorine}}\right) 100 \text{ g of chlorine} = 57.3 \text{ g of calcium}$$

From these calculations we can see that we don't have enough calcium to react with 100 g of chlorine, but we do have enough chlorine to react with 50.0 g of calcium. The calcium will be used up along with 87.3 g of chlorine, and 12.7 g of chlorine will remain.

Chapter 2

S1.   Since the attraction of the moon (Tranquility Base) is 1/6 of the earth, the earth weight should be divided by 6.

$$165 \text{ lbs}/6 = 27.5 \text{ lbs or } 75 \text{ kg}/6 = 12.5 \text{ kg}$$

S2.   If a rock on the moon (Tranquility Base) weighs one sixth as on the earth, to find its weight on earth, multiply by six; if a rock weighs 1.32 times more on Saturn, multiply the earth weight by 1.32.

$$5.7 \text{ kg x } 1.32 = 7.5 \text{ kg}$$

S3.   To work this problem, you must remember that the chemical symbol tells the number of protons in the atom, and the number of protons equals the atomic number, so refer to the periodic table.

a. C stands for carbon, atomic number 6 in the periodic table; since the mass number is given as 12 (the number above and to the left of the C), the number of neutrons is 12 - 6 = 6.

b. Referring to the periodic table for U gives an atomic number of 92, which is also specified in the symbol; since 235 is the mass number, the number of neutrons must be 235 - 92 = 143 neutrons.

c. C is the symbol for the atom (in this case carbon, at. no. = 6) and the mass number 14; thus the number of neutrons is 14 - 6 = 8.

d. As in b. 7 is the proton number (and the symbol points to N, #7 in the periodic table), and the mass number is 14, so the number of neutrons is 7.

S4.   We are given the ratio of hydrogen atoms to oxygen atoms, and the weights of each of these (look at the symbols in Exercise 22 in ECOT). Since the weight of hydrogen is 1 amu (and there are two hydrogens to each oxygen) and the weight of oxygen is 16 amu, the weight of a unit of water must be 18 amu. So what % is oxygen of the total weight?

$$\% \text{ oxygen } = \left(\frac{\text{weight of oxygen}}{\text{weight of oxygen} + \text{hydrogen}} \right) \times 100\%$$

$$= \left(\frac{16 \text{ amu}}{16 \text{ amu} + 2 \text{ amu}} \right) \times 100\%$$

$$= 89\%$$

The answer can be confirmed by subtracting the percentage of hydrogen stated in Exercise #22 of ECOT: $100\% - 11\% = 89\%$.

S5.   This problem is very similar to Exercise #24 in ECOT,

Assume the atoms are lined up with their diameters touching as we did for the gold atoms in #20 in ECOT. How many atoms will fit on the side of a 1 cm cube?

$$1 \text{ cm} \left(\frac{1 \text{ atom}}{2.9 \times 10^{-8} \text{ cm}} \right) = 3.4 \times 10^{7} \text{ atoms}$$

Since there are the same number of atoms on each edge of the cube, the number of atoms in the 1 cm cube will be

$$\left(3.4 \times 10^{7} \right)^{3} = 3.9 \times 10^{22} \text{ atoms/cm}^{3}$$

Since we know the weight of one atom, we can multiply that by the number of atoms to get the total weight:

$$\frac{3.9 \times 10^{22} \text{ atoms}}{\text{cm}^{3}} \left(\frac{3.42 \times 10^{-22} \text{ g}}{\text{atom}} \right) = 13.3 \text{ g/cm}^{3}$$

This value compares well with the measured density of 13.59 g/cm^3. What does this imply about the model we have chosen for the way in which the atoms are packed in the cube?

S6.   The mixture contains more of the isotope which more heavily contributes to the average atomic weight. Since the average is almost 7, ^{7}Li must be present in much greater percentage than ^{6}Li.

Chapter 3

S1.   Let y = fraction of mass 10 in mixture. Then, $1 - y$ = fraction of mass 11 in mixture.

Each isotope contributes to the *weighted average* of the naturally occurring mixture by its *weight* and the *fraction* in which it is present. Thus, if ^{10}B is present as fraction y, then its contribution to the mixture will be $10y$.

Using the numbers:
$$10y + 11(1 - y) = 10.81$$
$$y = 0.19$$

159

Since we let y = the fraction of mass 10 in the mixture, multiplying 0.19 by 100 gives us the percentage of ^{10}B in the mixture, or 19%; thus, ^{11}B is 81%.

S2.   This reaction is analogous to the reaction producing NaCl in Fig. 3.3. Use Lewis structures to represent the valence electrons.

$$K\cdot \ + \ \cdot \ddot{\underset{\cdot\cdot}{F}} : \longrightarrow \ K^{+} \ + \ :\ddot{\underset{\cdot\cdot}{F}} :^{-}$$

The equation for the reaction in elemental symbols is $K + F \rightarrow K^{+} + F^{-}$.

S3.   The secret here is to pick the *ionic* compounds, those made from a *metal* and a *nonmetal*, one atom from the left part of the periodic table and one atom from the right; if soluble in water, the solution will conduct. KBr is a good candidate, but CH_4 is not. Those that conduct are a, d, f.

S4.   If the compound is made up of a metal and a nonmetal, the name of the metal comes *first*. a. lithium oxide; b. propane; c. aluminum sulfide; d. hydrogen sulfide; e. potassium permanganate; f. nitrogen oxide (or dioxide).

S5.

Ethane, C_2H_6, has 7 covalent bonds, 7 bonding pairs of electrons, and 18 total electrons (each C has 6 electrons, each H has 1).

S6.   If you can calculate the total number of negative charges, you know the charges on the cations, because the total charge on a compound must be zero. a. Each F^- has a -1 charge, so Ba must be Ba^{2+}; b. V^{5+}; c. Os^{8+}; d. W^{2+}.

S7.   If you can calculate the total number of positive charges, you know the charges on the anions, because the total charge on a compound must be *zero*. a. Each Na^+ has +1 and there are two of them resulting in +2, so Se must be Se^{2-}; b. Br^-; c. N^{3-}; d. C^{4-}.

S8.   The naturally occurring mixture is composed of two different isotopes, each contributing to the total. If the fraction of Cl-35 is 0.758, the fraction of the other is 0.242.

$$\text{fract}(35) \times \text{mass}(35) + \text{fract}(m) \times \text{mass}(m) = 35.45,$$
where m is the mass of the unknown isotope.

$$\text{or} \quad 0.758(35) + .242(m) = 35.45$$
$$m = 36.86 \text{ or } 37$$

S9. $0.722(85) + 0.278(87) = 85.56$ amu

Chapter 4

S1. a. From the periodic table, uranium, U, has 92 p, 146 n, 92 e; b. 9 p, 10 n, and, because this is the fluoride ion with a charge of -1, it has 10 e; c. 3p, 3 n, 0 e; d. 54 p, 78 n, 54 e.

S2.

a. $^{20}_{9}F \longrightarrow \ ^{20}_{10}Ne + \beta$

b. $^{168}_{70}Yb \longrightarrow \ ^{168}_{70}Yb + \gamma$

c. $^{2}_{1}H + \ ^{13}_{7}N \longrightarrow \ ^{1}_{1+}p + \ ^{14}_{7}N$

d. $^{206}_{83}Bi \longrightarrow \ ^{0}_{1}n + \ ^{206}_{82}Pb$

e. $^{210}_{83}Bi \xrightarrow{\beta} \ ^{210}_{84}Po$

S3.

actual mass of U-235 atom = 235.044
mass of neutron = <u>1.009</u>
total mass of reactants = 236.053 amu

actual mass of Br-89 = 88.89
actual mass of La-144 = 143.901
mass of 3 neutrons = <u>3.027</u>
total mass of products = 235.818 amu

loss of mass = 0.235 amu

S4. $^{238}_{92}U \xrightarrow{\alpha} \ ^{234}_{90}Th \xrightarrow{\beta} \ ^{234}_{91}Pa \xrightarrow{\beta} \ ^{234}_{92}U \xrightarrow{\alpha} \ ^{230}_{90}Th$

S5.

$^{232}_{90}Th \xrightarrow{\alpha} \ ^{228}_{88}Ra \xrightarrow{\beta} \ ^{228}_{89}Ac \xrightarrow{\beta} \ ^{228}_{90}Th \xrightarrow{\alpha} \ ^{224}_{88}Ra \xrightarrow{\alpha} \ ^{220}_{86}Rn$

$^{220}_{86}Rn \xrightarrow{\alpha} \ ^{216}_{84}Po \xrightarrow{\alpha} \ ^{212}_{82}Pb \xrightarrow{\beta} \ ^{212}_{83}Bi \xrightarrow{\beta} \ ^{212}_{84}Po \xrightarrow{\alpha} \ ^{208}_{82}Pb$

S6. Calcium is the biologically interesting element. What do you think happens when the body absorbs Ra instead of Ca? What parts of the body will be affected first?

Chapter 5

S1. $^{112}_{48}\text{Cd} + ^{1}_{0}\text{n} \rightarrow ^{113}_{48}\text{Cd}$

S2.

$^{232}_{90}\text{Th} + ^{1}_{0}\text{n} \rightarrow ^{233}_{90}\text{Th}$

$^{233}_{90}\text{Th} \rightarrow ^{233}_{92}\text{U} + 2 \, ^{0}_{-1}\text{e}$

S3. $^{23}_{11}\text{Na} \rightarrow ^{23}_{10}\text{Ne} + ^{0}_{+1}\text{e}$

First half-life results in 50% of the original; second half life results in 25% of original. So, since the sample decayed to 25% of the original, it passed two half-lives in 62 months. The half-life is therefore 31 months.

S4. $^{198}_{80}\text{Hg} + ^{1}_{0}\text{n} \rightarrow ^{199}_{79}\text{Au} + ^{0}_{+1}\text{e}$

The second part of the question is of the "Think, Speculate, . . ." variety.

S5. Both iodide and chloride in the halogen family (Group 17 or VIIA) are important in the chemistry of life. Cesium is a member of the alkali family, Group 1, with sodium and potassium.

I-131 has a half-life of 8 days (Table 5.3 in ECOT) and Cs-137 a half-life of 30 years. Using the rule of thumb of waiting ten half-lives for radiation to become tolerable, the Ukrainians would have to wait 300 years for the Cs (the isotope with the longest half-life) to die out.

S6. $^{233}_{92}\text{U} + ^{1}_{0}\text{n} \rightarrow ^{160}_{62}\text{Sm} + ^{70}_{30}\text{Zn} + 3\,^{1}_{0}\text{n}$

S7. Don't forget to *subtract* the background reading from the sample reading.

The ratio of the two readings is (35-6)/(42-6) or 0.80. This means that 80% of the radiating material is left. Using Fig. 5.3 in ECOT, find the value of the fraction of half-life expired (about 0.3 half-life).

The half-life of C-14 is 5730 years, so (0.3)(5730) = 1719 yrs. Since the artifact may have been in existence around the time in question (approx. 200 C. E.), it may be from the Mithras cult. What other factors might enter into the decision? What other tests might be useful?

S8.   a. catching the fish; b. cleaning the fish.

Chapter 6

S1. Recall that an alkane is C_nH_{2n+2}, and an alkene (one double bond) is C_nH_{2n}. If a formula has fewer than 2n hydrogens, it has more than one double bond. (Ignore triple bonds, since the question is about double bonds.) Choices *c* and *f* each have more than one double bond.

S2. *c*.

S3. Write the molecular formulas of each of these, and you will see that *a* and *c* are isomers.

S4. *a* and *b* are aromatic hydrocarbons.

S5. Again, write molecular formulas. *a* and *b* are isomers.

S6. Whenever there is a double bond, one mole of H_2 can be added; a triple bond adds two moles of H_2. Sketch expanded structures for *a* and *b* to determine the presence or absence of double bonds.
a. 1 mole; b. none; c. 2 moles; c. 3 moles.

S7. a. 1-pentene; b. 2-pentene; c. 2-pentene;   d. 1-pentene.

S8. Follow the rules you have learned and make the models indicated. Check them with your study group.

Chapter 7

S1. vol A = 1250 cc, vol B = 120 cc

$$\left(\frac{\text{vol A}}{\text{vol B}}\right) = \left(\frac{1250 \text{ cc}}{120 \text{ cc}}\right) = 10.4$$

S2. a. Use the density of TEL , the volume of a teaspoonful, and the fact that the TEL in the teaspoon is 64% Pb to calculate the amount of Pb in a gallon of gasoline.

$$\left(\frac{3.5 \text{ mL TEL}}{\text{gal of gasoline}}\right)\left(\frac{1.653 \text{ g Pb}}{\text{mL TEL}}\right) 0.64 = 3.7 \text{ g Pb/gal of gasoline}$$

$$\left(\frac{3.7 \text{ g Pb}}{\text{gal}}\right)\left(\frac{125 \times 10^9 \text{ gal}}{\text{yr}}\right) = 462 \times 10^9 \text{ g Pb/yr} = 4.6 \times 10^{11} \text{ g Pb/yr}$$

b. Calculating the number of grams of Pb/day:

$$\left(\frac{4.6 \times 10^{11} \text{ g Pb}}{365 \text{ days}}\right) = 1.3 \times 10^9 \text{ g/day}$$

c. For each day:

$$\left(\frac{1.3 \times 10^9 \text{ g Pb}}{248 \times 10^6 \text{ persons}}\right) = 5.1 \text{ g of Pb/person each day}$$

S3.

$$CH_3 - \underset{\underset{CH_3}{|}}{\overset{\overset{CH_3}{|}}{C}} - \underset{\underset{CH_3}{|}}{\overset{\overset{CH_3}{|}}{C}} - H$$

S4.

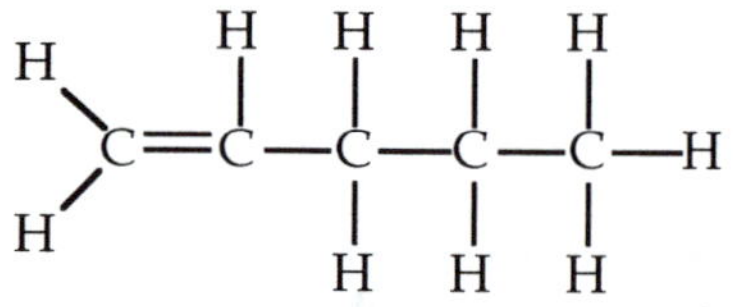

1-pentene

S5. Pentane is a likely product.

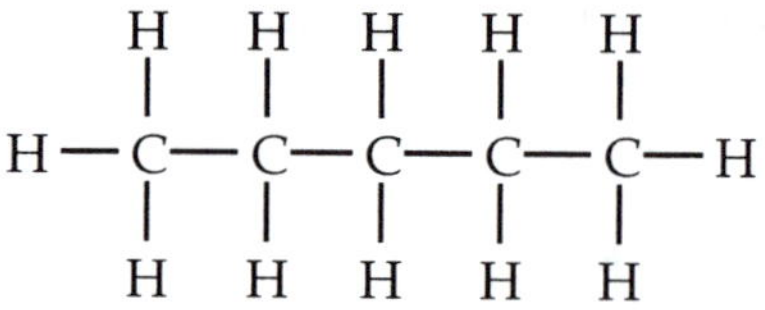

S6. C_9H_{20}

Chapter 8

S1. Draw the structure of nonane, and determine that there are 8 C-C bonds and 20 C-H bonds. Use the estimating method: 28 bonds x 53.2 kcal/bond = 1490 or 1500 kcal.

S2. We can solve this intuitively. We have equal quantities of water at temperatures 30 °C apart. The final temperature will be 15° from each initial temperature, or 45 °C.

S3. How many calories are contributed by each type of nutrient? Assume 100 g of food.

37 g fat x 9 cal/g fat = 333 cal; 45 g protein x 4 cal/g protein = 180 cal; 18 g carbo x 4 cal/g carbo = 72 cal. Add to get total calories or 585 cal.

$$\% \text{ cal from fat} = \frac{333 \text{ cal fat}}{585 \text{ cal total}} \times 100$$
$$= 57\%$$

S4. From table 8.3,

pizza	180 cal	35 min
malt	502 cal	97 min
	total	133 min (2.2 hrs)

S.5

$$3.5 \text{ kg}\left(\frac{1000 \text{ g}}{\text{kg}}\right)\left(\frac{1 \text{ cal}}{\text{g }°\text{C}}\right)(28 \,°\text{C})\left(\frac{4.2 \text{ joules}}{\text{cal}}\right)\left(\frac{\text{watt x second}}{\text{joule}}\right)\left(\frac{1}{250 \text{ watts}}\right)\left(\frac{1 \text{ hr}}{3600 \text{ sec}}\right) = 0.46 \text{ hr}$$

Chapter 9

S1.   a. $2\ H_2O_2 \rightarrow 2\ H_2O + O_2$

    b. $2\ Cs + Cl_2 \rightarrow 2\ CsCl$

    c. $4\ Al + 3\ O_2 \rightarrow 2\ Al_2O_3$

    d. $H_2SO_4 + 2\ NaOH \rightarrow Na_2SO_4 + 2\ H_2O$

    e. $CuSO_4 + 2\ NaOH \rightarrow Cu(OH)_2 + Na_2SO_4$

    f. $C_3H_8 + 5\ O_2 \rightarrow 3\ CO_2 + 4\ H_2O$

    g. $2\ C_8H_{18} + 25\ O_2 \rightarrow 16\ CO_2 + 18\ H_2O$

S2. $2\ CH_4 + 3\ O_2 \rightarrow 4\ H_2O + 2\ CO$

$$1000\ g\ O_2 \left(\frac{1\ mol\ O_2}{32\ g\ O_2}\right)\left(\frac{2\ mol\ CO}{3\ mol\ O_2}\right)\left(\frac{28\ g\ CO}{1\ mol\ CO}\right) = 583\ g\ CO$$

S3.

$$\left(\frac{68\ g\ HNO_3}{100\ g\ solution}\right)\left(\frac{mol\ HNO_3}{63\ g\ HNO_3}\right)\left(\frac{1.41\ g\ solution}{mL\ of\ solution}\right)\left(\frac{1000\ mL\ solution}{L\ solution}\right) = 15.2\left(\frac{mol\ HNO_3}{L\ solution}\right)$$

$$= 15.2\ M$$

S4.   In a dilution the number of moles in the first solution = the number of moles in the diluted solution:

$$5\ L\ HNO_3 \left(\frac{6\ mol}{L\ HNO_3}\right) = 30\ mol\ HNO_3$$

$$V = 30\ mol\ HNO_3 \left(\frac{L}{15.2\ mol\ HNO_3}\right) = 1.97\ L = 2\ L$$

S5.

    $0.00015\ g/L = 0.15\ mg/L$

    If 1 mL of blood weighs 1.05 g, 1L of blood weighs 1050 g or 1.05 kg

    1 ppm = 1 mg/kg, so $\left(\dfrac{0.15\ mg/L}{1.05\ kg/L}\right) = 0.14\ ppm$

S6.   Let x be the amount of sugar which must be added; set up the equation for calculating % concentration with percent = 5, add x in both the numerator and denominator, and solve for x.

$$5 = \left(\frac{17.5\text{ g} + x}{500\text{g} + 17.5\text{ g} + x}\right) \times 100$$

$$\left(17.5 + x\right)100 = 5\left(517.5 + x\right)$$

$$x = 8.8\text{ g}$$

Chapter 10

S1. a. $CaO + H_2O \rightarrow Ca(OH)_2$

  b. $MgO + H_2O \rightarrow Mg(OH)_2$

S2. a. $KOH + HBr \rightarrow KBr + H_2O$

  b. $2\,KOH + H_2SO_3 \rightarrow K_2SO_3 + 2\,H_2O$

  c. $Ca(OH)_2 + 2\,HF \rightarrow CaF_2 + 2\,H_2O$

  d. $3\,Sr(OH)_2 + 2\,H_3PO_4 \rightarrow Sr_3(PO_4)_2 + 6\,H_2O$

S3. a. $[H_3O^+] = 0.001$ M; pH = 3;

  b. pOH = 3; pH = 11;

  c. pH = 2; pOH = 12.

S4. a. pH = 1;

  b. pH = 10;

  c. If pOH = 3, pH = 11; if pOH = 5, pH = 9. The solution with pH = 9 has the greater $[H_3O^+]$.

  d. If pOH = 10, pH = 4; if pOH = 12, pH = 2. The solution with pH = 2 has the greater $[H_3O^+]$.

S5. a. Always start with a *balanced* equation! $Ca(OH)_2 + 2\,HCl \rightarrow CaCl_2 + 2\,H_2O$

$$0.250\text{ mL}\left(\frac{0.1\text{ mol}}{L}\right) = 0.0250\text{ mole Ca (OH)}_2$$

From the balanced equation,1 mole of Ca $(OH)_2$ requires 2 moles of HCl.

$$0.0250\text{ mol Ca (OH)}_2\left(\frac{2\text{ mol HCl}}{1\text{ mol Ca (OH)}_2}\right)\left(\frac{1\text{ L HCl}}{0.2\text{ mol HCl}}\right) = 0.250\text{ L HCl or 250 mL HCl}$$

  b. In water, CaO becomes $Ca(OH)_2$, and the neutralization equation is the one in *a*. above.
(See S1. above.)

$$3.7 \text{ g Ca O} \left(\frac{1 \text{ mol Ca O}}{56 \text{ g}} \right) = 0.066 \text{ mol Ca O}$$

1 mole CaO produces 1 mole Ca(OH)$_2$ in solution, and the reaction proceeds as above.

$$0.066 \text{ mol Ca (OH)}_2 \left(\frac{2 \text{ mol HCl}}{1 \text{ mol Ca (OH)}_2} \right) \left(\frac{1 \text{ L}}{0.2 \text{ mol HCl}} \right) = 0.66 \text{ L} = 660 \text{ mL HCl}$$

c. An acid cannot neutralize an acid.

Chapter 11

S1. a. $Pb \rightarrow Pb^{2+} + 2 \text{ e}^-$
                  $2 \text{ H}^+ + 2 \text{ e}^- \rightarrow \text{ H}_2$

b. overall: $Pb + 2 \text{ H}^+ \rightarrow Pb^{2+} + \text{ H}_2$

Since lead is oxidized in the presence of the hydrogen ion, or, to put it another way, since hydrogen ion oxidizes lead, the lead half-cell reaction must be above the hydrogen half-cell reaction in the Standard Reduction Table.

S2. a. No. The large positive voltage means that the Au^{3+} ion has a strong tendency to acquire electrons, thus being reduced. For gold to dissolve in acid it would have to be oxidized.

b. Nothing. Au^{3+} will be reduced in the presence of Fe, not the other way around.

S3. Write the half-cell reactions in this order:

$Sr^{2+} + 2e \rightarrow Sr$	-2.89v
$Hg^{2+} + 2e \rightarrow Hg$	+0.85v
$Ag^{2+} + e \rightarrow Ag^+$	+1.98v

a. Now you can see that since the strongest oxidizing agent is the most easily reduced, the answer is Ag^{2+}.

b. The strongest reducing agent is the most easily oxidized, or Sr.

c. No, since Hg^{2+} will be easily reduced, and therefore Hg won't be oxidized in the presence of Sr^{2+}. In the same way, Sr will be oxidized, so Sr^{2+} won't be reduced.

S4. $Cd \rightarrow Cd^{2+} + 2 \text{ e}^-$ +0.40 v
              $2H^+ + 2 \text{ e}^- \rightarrow \text{ H}_2$ 0 v

overall: $Cd + 2H^+ \rightarrow Cd^{2+} + \text{ H}_2$ +0.40 v

S5. $2(Ni \rightarrow Ni^{2+} + 2 \text{ e}^-)$ +0.26 v
              $O_2 + 4H^+ + 4e^- \rightarrow 2 \text{ H}_2O$ +1.23 v

overall: $2 Ni + O_2 + 4 \text{ H}^+ \rightarrow Ni^{2+} + 2 \text{ H}_2O$ +1.49 v

Note that, although the Ni half-cell reaction must be multiplied by 2 to balance the equation, the voltage is *not* doubled. Why?

S6. Cd^{2+}, perhaps from $Cd(NO_3)_2$ or $CdCl_2$.

Chapter 12

S1.   Pressure is doubled, so volume is halved. The new volume = 1 L.

S2. $V_1 = 25$ L $V_2 = ?$
       $P_1 = 760$ mm-Hg $P_2 = 350$ mm-Hg
       $T_1 = 25$ °C or 298 K $T_2 = 100$ °C or 373 K

$$V_2 = 25 \text{ L} \left(\frac{760 \text{ mm - Hg}}{350 \text{ mm - Hg}} \right) \left(\frac{373 \text{ K}}{298 \text{ K}} \right) = 68 \text{ L}$$

S3. $V_1 = 5$L $V_2 = 10$ L
       $P_1 = 760$ mm-Hg $P_2 = ?$
       $T_1 = 25$ °C or 298 K $T_2 = -10$ °C or 263 K

$$P_2 = 760 \left(\frac{5 \text{ L}}{10 \text{ L}} \right) \left(\frac{263 \text{ K}}{298 \text{ K}} \right) = 335 \text{ mm- Hg}$$

S4. $V_1 = 250$ L $V_2 = ?$
       $P_1 = 1$ atm $= 760$ mm-Hg $P_2 = 75$ mm-Hg
       $T_1 = 25$ °C or 298 K $T_2 = -53$ °C or 220 K

$$V_2 = 250 \text{ L} \left(\frac{760 \text{ mm - Hg}}{75 \text{ mm - Hg}} \right) \left(\frac{220 \text{ K}}{298 \text{ K}} \right) = 1870 \text{ L}$$

S5.   When it registers "empty" on the gauge, it's not empty. The pressure in the tank is equal to atmospheric pressure, or 14.7 psi.

S6. $H_2 + Cl_2 \rightarrow 2 \text{ HCl}$

According to Avogadro, this equation can be interpreted to read "hydrogen and chlorine react in a 1:1 volume ratio to produce 2 volumes of HCl." Therefore 6 L of hydrogen will react with 6 L of chlorine to produce 12 L of HCl. If we start with 12 L of hydrogen and 6 L of chlorine, 6 L of hydrogen will be used and 6 L of hydrogen will remain unreacted. In the vessel after the reaction will be 12 L of HCl and 6 L of H_2.

Chapter 13

S1.   Weigh the mineral and the saucer. Place the glass on the saucer, and fill it with water until it almost overflows. Carefully put the mineral sample in the water, catching the displaced water in the saucer. Reweigh the saucer plus water. Since 1 g of water = 1 mL of water, the volume of the displaced water is now known and is equal to the volume of the mineral. Calculate the density of the mineral.

S2. If a wood's density is greater than water's, it will sink. So you can calculate the density of each of the woods and compare it with that of water. Another way is to find the mass which, when used with a volume of (2 cm x 2 cm x 2 cm) or 8 cm³, will give a density less than or equal to 1 g/cm³. Then any wood with a mass less than that will float.

$$m = V \times d = 8 \text{ cm}^3 \left(\frac{1 \text{ g}}{\text{cm}^3} \right) = 8 \text{ g}$$

So, the wood sample weighing less than 8 g will float.

S3. a. propyl acetate; b. propyl stearate.

S4. a. methyl benzoate; b. propyl butyrate; c. octyl propionate; d. butyl propionate; c. capryl oleate.

S5. The Ca^{2+} ion will precipitate with the anion of a soap but not with the anion of a detergent.

Chapter 14

S1. a. since 1 mg/L = 1 ppm, 20 ppm = 20 mg/L.

b. 20 ppm = 20 mg/L = 20 mg/kg. $20 \dfrac{\text{mg}}{\text{kg}} \left(\dfrac{1 \text{ kg}}{10^6 \text{mg}} \right) \times 100 = 0.020\%$

c. $400 \dfrac{\text{mg}}{\text{L}} \left(\dfrac{1 \text{ g}}{1000 \text{ mg}} \right) \left(\dfrac{1 \text{ mol}}{75 \text{ g}} \right) = 5.3 \times 10^{-3} \, M$

d. $0.025 \dfrac{\text{mol}}{\text{L}} \left(\dfrac{40 \text{ g}}{\text{mol}} \right) \left(\dfrac{1000 \text{ mg}}{\text{g}} \right) = 1000 \text{ mg/L} = 1000 \text{ ppm}$

e. $2 \text{ ppm} \left(\dfrac{10^3 \text{ ppb}}{\text{ppm}} \right) = 2000 \text{ ppb}$

S2. Industrial emissions produce the highest percentage of VOCs.

S3. a. $SO_2 + Ca(OH)_2 \rightarrow CaSO_3 + H_2O$

b. $CO_2 + H_2O \rightarrow H_2CO_3$

c. $SO_3 + H_2O \rightarrow H_2SO_4$

d. $O_2 + O \rightarrow O_3$

e. $FeS_2 + 2 O_2 \rightarrow Fe + 2 SO_2$

S4. From Sec. 7.1, there are 200 million cars in the U.S., or 200 million people driving (assuming 1 per person):

$$200 \times 10^6 \text{ people} (0.01) \left(\frac{52 \text{ candy wrappers}}{\text{yr}} \right) 10 \text{ yrs} = 1.04 \times 10^9 \text{ or } 1{,}040{,}000{,}000 \text{ candy wrappers!}$$

S5. a. primary; b. neither; c. secondary; d. neither; e. secondary.

S6. A good "Think, Speculate . . ." question. Refer to Sec. 10.6 in ECOT and, in your thinking, distinguish between rain that contains oxides from natural sources, such as CO_2, SO_2 from volcanic action, etc., from acid rain produced from human pollution.

Chapter 15

S1. $CH_3(CH_2)_7 - CH = CH - (CH_2)_7 - CO_2H + H_2 \rightarrow CH_3(CH_2)_{16} - CO_2H$

S2. $100 \text{ g oleic acid} \left(\frac{1 \text{ mol oleic acid}}{282 \text{ g}} \right) \left(\frac{1 \text{ mol } H_2}{\text{mol oleic acid}} \right) \left(\frac{2.0 \text{ g } H_2}{\text{mol } H_2} \right) = 0.71 \text{ g}$

S3. $CH_3 - (CH_2)_7 - CH = CH - (CH_2)_7 - CO_2H + I_2 \rightarrow$

$$CH_3 - (CH_2)_7 - \underset{\underset{I}{|}}{CH} - \underset{\underset{I}{|}}{CH} - (CH_2)_7 - CO_2H$$

S4. $100 \text{g} \left(\frac{1 \text{ mol}}{282 \text{ g}} \right) \left(\frac{1 \text{ mol } I_2}{\text{mol oleic acid}} \right) \left(\frac{254 \text{ g } I_2}{\text{mol } I_2} \right) = 90.1 \text{ g}$

S5. The animal fat with the highest percentage of monounsaturated fatty acids is human fat (52%). The vegetable oil with the highest percentage of monounsaturated fatty acids is olive oil (85%).

The animal fat with the lowest percentage of saturated fatty acids is human fat (35%). The vegetable oil with the lowest percentage of saturated fatty acids is tung oil (5%).

Chapter 16

S1. a. yes; b. no; c. yes; d. no; e. no (Sec. 16.10 of ECOT points out that anything that is planar is superposable on its own image.)

S2. a. glucose; b. maltose; c. triglycerides; d. cellobiose; e. lactose.

S3. a. yes; b. yes; c. yes; d. no; e. yes.

S4. a. ketopentose; b. aldohexose; c. aldopentose; d. ketohexose.

S5. a. 2; b. 4; c. 3; d. 3.

S6. a. $2^2 = 4$; b. $2^3 = 8$.

S7. There are 4 cal/g available from any *digestible* carbohydrate. Are all of the examples digestible?

Chapter 17

S1.

a.

b.

S2.

a.

b.

S3. Student project.

S4. No. Sickle-cell anemia is a disease caused by an error in genetic information. Each time another hemoglobin molecule is manufactured, the error is repeated.

S5.

L-alanine   D-alanine
(from Sec. 17.3
   in ECOT)

Chapter 18

S1. 1st position: 4 possible bases
     2nd position: 4 possible bases
     3rd position: 4 possible bases

     thus, 4 x 4 x 4 = 64 possible codons

S2. 1st position: 4 possible bases
     2nd position: 4 possible bases

     Thus, 4 x 4 = 16 possible codons

     No, a minimum of twenty codons are required to give each amino acid in Table 18.1 a unique codon.

S3. adenine: 21%   thymine: 21%
     cytosine: 29%   guanine: 29%

S4.

     GGUACGU – AUG   – GAA – GUC   – UGG   – AAA   – UUU   – UAG   – CUCUAGC
                (start)   (stop)
                Met   – Glu   – Val   – Trp   – Lys   – Phe

S5. $\dfrac{1 \text{ success}}{278 \text{ attempts}}$ x 100 = 0.36%

Chapter 19

S1. It has 2 chiral carbons (See Section 16.10); therefore, it can have 2^n or 4 enantiomers, where n = the number of chiral carbons.

S2. From the formula, calculate the number of grams of calcium and phosphorus in a mole of hydroxyapatite.

     Ca: 40 g/mol x 10 mol = 400 g
     P: 31 g/mol x 6 mol = 186 g
     Since the amount of P is just about half the amount of Ca, the statement is true.

S3.   Student answer.

S4.   RDA for Ca for a young male is 1200 mg. Half would be 600 mg. One cucumber contains 37 mg of Ca.

a. 600 mg (1 cuke/37 mg) = 16.2 cukes;

b. $\left(\dfrac{37 \text{ mg}}{34 \text{ oz cuke}}\right)\left(\dfrac{1 \text{ oz}}{28.4 \text{ g}}\right)\left(\dfrac{1 \text{ g}}{1000 \text{ mg}}\right) \times 100 = 0.004\%$

S5.   RDA for thiamine is 1.1 mg; 0.25 x 1.1 mg = 0.275 mg. If one onion contains 0.2 mg, then 1.4 onion contains 0.275 mg of thiamine.

S6. $\left(\dfrac{0.2 \text{ mg}}{110 \text{ g}}\right)\left(\dfrac{1 \text{ g}}{1000 \text{ mg}}\right) \times 100 = 1.8 \times 10^{-4}\%$

S7.   a.   The RDA for vitamin A for an adult female is 800 RE. Since 1 RE is 3.33 IU, the RDA in IU would be 2664. 14,580 - 2664 = 11, 916 or 11, 920 IUs in excess. b. The excess will be stored in the liver.

S8.   a.   The person ingests 227 mg of vitamin C. Since the RDA is 60 mg, this is an excess of 167 mg.
b.   This amount will last about a day;
c.   The excess will be excreted in the urine.

Chapter 20

S1.   To fit in the table, μg must be converted to mg.

$\left(\dfrac{10 \text{ μg}}{\text{kg}}\right)\left(\dfrac{1 \text{ mg}}{1000 \text{ μg}}\right) = 1 \times 10^{-2}$ mg/kg. Tetrodotoxin would fit between TCDD and muscarine.

S2.   a.   The LD_{50} for acetaminophen is 0.34 g/kg or 340 mg/kg. For an 80 kg person to exceed the LD_{50}, he or she would have to ingest 340 mg/kg (80 kg) or 27,200 mg of the compound. Since there are 300 mg of acetaminophen/tablet, the person would have to take 27,200 mg/300 mg/tablet or 90.7 tablets.

b.   90.7 tablets $\left(\dfrac{1 \text{ gulp}}{3 \text{ tablets}}\right)\left(\dfrac{200 \text{ mL}}{\text{gulp}}\right)\left(\dfrac{1 \text{ L}}{1000 \text{ mL}}\right) = 6 \text{ L}$

S3. $\left(\dfrac{0.65 \text{ g}}{\text{kg}}\right)\left(\dfrac{0.225 \text{ kg}}{\text{rat}}\right) = 0.15 \text{ g/rat}$

S4.   a. $\left(\dfrac{2.5 \times 10^{-4} \text{ g}}{0.250 \text{ kg}}\right)\left(\dfrac{1000 \text{ mg}}{\text{g}}\right) = 1 \text{ mg/kg}$

b. between tubocurarine chloride and rotenone

c. At the rate of 80 mg/80 kg human, 1 g or 1000 mg would be lethal to half a population of 12 people.

Chapter 21

S1. If you draw a large enough network it becomes easy to see that the ratio of Si to O is really 1:2, not 1:4. However, be aware that the apparently planar structure drawn below is really 3-dimensional. The representation in Figure 21.25 in ECOT better illustrates that 3-dimensionality.

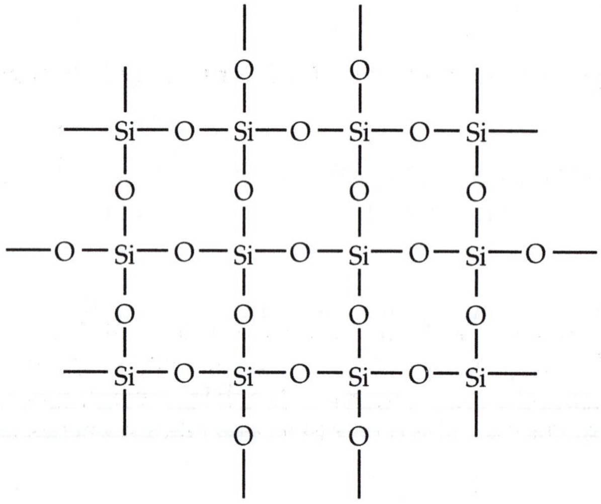

```
              |        |
     |        O        O        |
  — Si — O — Si — O — Si — O — Si —
     |        |        |        |
     O        O        O        O
     |        |        |        |
  — O — Si — O — Si — O — Si — O — Si — O —
     |        |        |        |
     O        O        O        O
     |        |        |        |
  — Si — O — Si — O — Si — O — Si —
     |        |        |        |
              O        O
              |        |
```

S2. Use Figure 21.15 to help you. Identify the 1,6-diaminohexane portion of the polymer. Now change that part of the polymer to 1,4-diaminobutane—in other words, use 4 carbons instead of 6.

S3. a. CH_2=CH–CH_3
     b. Addition polymer

Chapter 22

S1. Sulfonate dyes contain sodium salts of sulfonic acid, and the presence of these sulfonate groups ($-SO_3^-$) on the large molecule make the molecule ionic and soluble in water. This alone does not account for their impermanence: they are also very large molecules that cannot penetrate into the cortex of the hair, and so are easily removed in washing.

S2. The semipermanent dyes, though somewhat water soluble, are small molecules that can penetrate the cortex and thus resist several washings. They diffuse slowly out of the cortex, however, and are not permanent.

S3. Since hair reflects light at a lower pH due to tight cuticle, rinsing with a mild acid solution, such as diluted vinegar or lemon juice, lowers the pH and achieves the desired lustrous appearance.

S4.	Although the density is given, we don't have to convert to mass/kg, since the problem asks for the *volume* of thioglycolic acid which would be required for a 50% chance of death in 100 52-kg individuals.

Volume required = 100 (52 kg x 0.15 mL/kg) = 780 mL

S5.	Civetone is the heavier molecule, and on that basis would be expected to be less volatile.

S6.	An SPF of 12 reduces UV exposure by a factor of 12. Two hours (120 min) in the sun using a sunscreen with an SPF of 12 would have the effect of 10 min in the sun with no protection.

Chapter 23

S1.	In scopolamine an ether linkage (—O—) exists between 2 carbons. This does not occur in atropine.

S2.	They are all local anesthetics containing amine groups.

S3.	Student project.

S4 – S6:
These three questions are best answered with an information table. Put *functional groups* across the top (S4), followed by *function* (S5) and *plant names* (S6), where appropriate. List all the compound names from the chapter down the left side. Make check marks for every functional group found in a compound, and write in function and plant names.

Tetrahedron Model

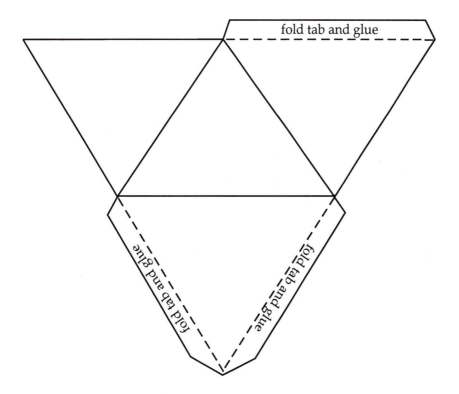

fold tab and glue

fold tab and glue

fold tab and glue

Cut out the 109° template at the right. Place it against toothpicks embedded in the gum drop to set the proper angle for bonds.

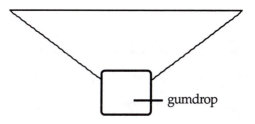

gumdrop

NOTES

NOTES

NOTES

NOTES

NOTES